SENT FORTH A DOVE

This is a presentation copy
individually numbered and
signed by the author.

Copy number 844 /1000

OTHER BOOKS BY JAMES HENDERSON:

Marooned
Rottnest: Links in History
Onward Boy Soldiers
Phantoms of the Tryall

SENT FORTH A DOVE

Discovery of the Duyfken

James Henderson

University of Western Australia Press

First published in 1999 by
University of Western Australia Press
Nedlands, Western Australia 6907

National Library of Australia
Cataloguing-in-Publication entry:

Henderson, James A., 1925– .
Sent forth a dove : discovery of the Duyfken.

Includes index.
ISBN 1 876268 24 7
ISBN 1 876268 25 5 (pbk.).

1. Duyfken (Ship). 2. Duyfken (Ship : Replica). 3. Sailing ships — Western Australia — Reproductions. 4. Australia — Discovery and exploration — Dutch. I. Title.

919.4041

Produced by Benchmark Publications Management, Melbourne
Consultant editor Bryce Moore, Perth
Design and typesetting by Allan Watson, Perth Editorial Service, Perth
Cover design by Derrick Stone, Melbourne
Maps by Adriaan de Jong
Typeset in 11 pt Bembo
Printed by PK Print, Perth

Sent Forth a Dove *is dedicated to Clive and Francis Yunkaporta, elders of Keerweer, who asked me to 'tell the true story for all Australians'.*

Foreword

IN THE seventeenth century, by their initiative and skill and use of the latest technology, the sailing ships of the Netherlands nation led the world. Thus equipped, Captain Willem Janszoon and the *Duyfken* discovered a vast new continent.

Fugro NV surveys today's wide world of land and sea, above and below, for the benefit of both our shareholders and our many and various customers.

In 1995 we felt that Australian author James Henderson of the *Duyfken* 1606 Replica Foundation deserved our support for his historical research initiatives for *Sent Forth a Dove*, and so it was that he took off for the remote Cape York Peninsula with a copy of the *Duyfken* 1605–06 discovery chart.

A couple of years later HRH Prince Willem shook hands with Henderson, our marine manager Grey Roughan and others at the laying of the keel of the *Duyfken* replica in Fremantle, Western Australia. We were actively involved in the research project when Grey Roughan accompanied the author on another expedition, to solve the 1618 mystery of the Dutch-named Willems River, which had long since vanished from Australian charts.

Fugro is proud to have been involved with the research for and publication of the book *Sent Forth a Dove*, the story of the Dutch discovery of Australia and the *Duyfken* 1606 Replica Project.

GERT-JAN KRAMER
President of the Fugro group of companies

FUGRO

Brief Glossary

APRON: A timber piece forming an inside stem of a ship, aft of the stempost.

BILGE: A part of the ship's bottom where it is nearly horizontal.

BLOCKS: Wooden pulley systems.

BOWSPRIT: A spar projecting forward from the bow to extend the spritsail and support the foremast.

BREASTHOOKS: Inside timbers in the bows, fastening the forepart of the structure.

CAULKING: The process of stopping up the seams in the planking with oakum or pitch.

COMPASS TIMBER: Naturally curved piece of timber.

DEAD-EYE: Round flat three-holed block for extending the shrouds.

FUTTOCKS: From 'foot hooks', a ship's middle timbers between the floor and top timbers.

GARBOARD STRAKE: The first range of planks laid on a ship's bottom next to the keel.

GUDGEON: Socket in which the rudder works.

HAWSEPIECES: Part of the bows with holes through which the anchor cable runs.

KEELSON: A heavy fore-and-aft centre-line beam extending from stem to sternpost, giving longitudinal strength.

KNEES: Angled grown timbers for supporting deck beams.

PINTLE: Pin or bolt on which the rudder turns.

PRAHU OR PRAO: A generic term referring generally to several types of swift Indonesian boats, some undecked, which are propelled by sails, oars or paddles.

RAHOUT: The fore-and-aft rail on the ship's sides, on which the yards are brought down.

SCUPPERS: Holes in the ship's side to shed water from the deck.

SHROUDS: A set of ropes forming part of the standing rigging, supporting the mast or topmast.

SPIRKETTING: Inside planking between the top of the waterways and the lower sills of portholes.

STRAKE: A continuous line of planking from stem to stern.

TIMBERS: Any of the curved pieces forming the ribs of a vessel.

TRANSOM: Horizontal beams across the stern.

TRUNNELS: From 'tree nails', wooden dowels for fastening the planking to the hull.

TUMBLEHOME: The inward incline of the ship's sides, above the widest part of the hull.

WATERWAYS: Moulded timbers forming a watertight joint between a ship's inboard and its deck, guiding the water toward the scuppers.

WALE: Broad thick timbers running along the ship's side.

WHIPSTAFF: Like a tiller extension, allowing the helmsman to operate the tiller from below the deck he stands on. The whipstaff passes through the deck in a bearing called a rowel.

WINDLASS: Winding device with a horizontal axle for hauling up the anchor.

YARDS: Spars slung across the mast to support the sails.

Contents

Acknowledgments

THIS BOOK *Sent Forth a Dove*, and the *Duyfken* 1606 replica undertaking, were integral parts of an enterprise which began in 1994, and thus a good number of the people involved with me in the initiation and development of the replica project also contributed in various ways to the story.

From the beginning of the dual process that we described as 'building our history', Dr Cornelis de Heer and Michael Young supported the work and gathered historical material and helpful people. Michael Young's enthusiasm was a key element in the success of the ambitious project.

When the stream of sixteenth- and seventeenth-century documents relating to the life and death of the 'Little Dove' began to emerge from Dutch archives, Adriaan de Jong translated and explained them for us with such skill that the ship and its era of Dutch discoveries came to life. Later he worked with the replica project's chief executive officer, Noel Robins, design coordinator Nick Burningham, master shipwright Bill Leonard and master rigger Igor Bjorksten to produce working drawings for the construction.

In the difficult task of describing the construction of the replica I was greatly assisted by contributions from Leonard, Burningham and de Jong. My fellow members of the *Duyfken* 1606 Replica Foundation encouraged the writing of the book, and the Foundation's public relations adviser Graham Cocks helped the book project wherever possible.

Marit van Huystee's research at the Western Australian Maritime Museum demonstrated for us beyond doubt, from sixteenth- and seventeenth-century Dutch shipping records, that our 1606 *Duyfken* was the second ship of the name. The award-winning Maritime Museum photographer Patrick Baker assisted with his professional skills.

The leading Australian cartographer W. A. R. Richardson, of South Australia, shared with me his extensive research findings, which dismissed claims that Spanish or Portuguese explorers had come to Australia before the *Duyfken*.

Historian Phillip Playford, who had assisted me in research for earlier maritime books, gave valuable advice about evidence of Dutch maritime history and, with Graeme Henderson, read and discussed my manuscript. At an early stage in my research, historian Lindsay Peet sent me some valuable references.

Editor Bryce Moore's skilful contribution assisted in clearly presenting the story of four centuries linking Europe, Asia and Australia.

Close to home, neighbour Roy Stone of Fremantle Sailing Club, winner of an impressive number of ocean races, added greatly to my slender navigational skills when we collaborated to produce a plan for a '*Duyfken* 1606 Rediscovery Expedition', using a yacht to sail from Banda to Cape York Peninsula in the wake of the 'Little Dove'. Club members Chris Hansen and Lyn and Mike Stewart were interested in the use of their vessels — respectively the fourteen-metre sloop *Sovereign* and the sixteen-metre yacht *Onfekomenadi* — but for various reasons we were unable to make either option viable.

After Bill Bunbury of ABC Radio's 'Talking History' series passed on to me what he had been told briefly about an alleged Aboriginal version of the events of 1606, the support of Fugro Survey Pty Ltd made possible a research expedition to Cape York Peninsula to reconstruct the events.

The then chairman of the Anti-discrimination and Reconciliation Commission, former High Court Justice Sir Ronald Wilson, kindly gave me written recommendations to assisted in communication with Aboriginal groups in remote places.

Ross Gardner, a senior civilian employee of the Navy at HMAS *Cairns* base, and Michael Fordham, now manager of Cairns regional office of the Queensland Government Department of the Premier and Cabinet, located the relevant Aboriginal elders, and their participation in my 1995 expedition made it a success.

The Aboriginal elders of Cape Keerweer, Yunkaporta brothers Clive and Francis, Aboriginal ranger Lawrie Booth of Mapoon and the Aborigines with whom I communicated at the centres of Napranam, Aurukun, Keerweer and Mapoon were all pleased to help. In Cairns, Aboriginal lawyer Noel Pearson of the Cape York Land Council agreed to discuss the traditional stories I had been given, and reminded me of what Captain James Cook had said of the Aborigines he encountered near Cooktown: '...an inoffensive race, no ways inclined to cruelty'.

Anthropologist Dr Peter Sutton of South Australia kindly discussed with me the traditional Aboriginal versions of the events of 1606. I had followed very humbly in some of his dedicated professional tracks of research.

Michael Pickering of Melbourne made available his Australian National University doctoral thesis, which dealt with unsupported allegations of Aboriginal cannibalism ('The bloodthirsty savages of popular myth are indeed myth, the products of fertile and foreign imaginations'); and Richard Buchhorn of the Catholic Social Justice Council sent me information from his studies of lurid accounts of Aboriginal cannibalism, in regard to which his persistent inquiries had found no substance.

Prof. Noel Loos of James Cook University, Queensland, referred me to his paper outlining the records of eight Dutch ships which sailed the Australian coasts and made contact with Aborigines between 1606 and 1756. Skippers were instructed to avoid molesting Aborigines, but also to seize some to bring back on their ships if possible!

The Royal Australian Navy increased my understanding by conducting a joint survey of the Australian discovery course of Willem Jansz of *Duyfken*. Lieutenant Commander John Maschke of the RAN Hydrographic Survey section's HMAS *Flinders* was extremely efficient, cooperative, helpful and hospitable in this operation of 'rediscovering Australia.'

J. A. H.

Introduction

THE WHOLE concept of the *Duyfken* is exciting. It has caught the imagination of West Australians as they have seen the vessel growing from a bare keel to a shapely little ship.

Many thousands of school children have been able to feel themselves part of the building process, watching the skilled artisans at work, even touching *Duyfken*'s timbers as they have been set in place. But *Duyfken* is more than simply a Dutch East India Company working ship: she has an important story.

Her voyage to the coast of the mysterious South Land needed to be studied and explained for all Australians. James Henderson has been much involved in sifting evidence and in tracking *Duyfken*'s voyages and her history. We are grateful to him for his commitment to the task and for this clear account of his researches.

Our *Duyfken*, the 'Little Dove', is a ship of friendship.

DR MICHAEL KAILIS
Chairman, *Duyfken* 1606 Replica Foundation

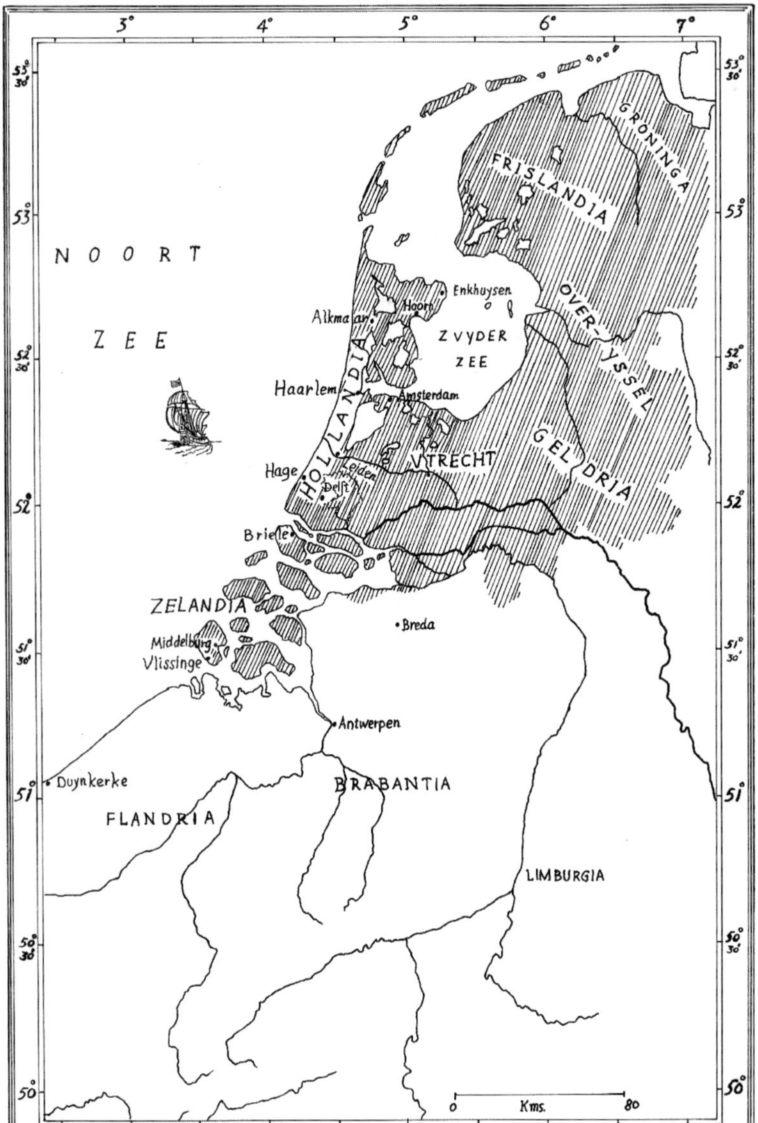

Historic map of the United Provinces, c.1600

Discovering *Duyfken*

THE FIRST European visit to Australia was that of the Dutch East India Company's vessel *Duyfken* ('Little Dove') in 1606. Historic documents referring to the voyage have existed in Dutch archives and published collections for centuries, forgotten or unpublicised until brought to light by our project to build an authentic replica of the tiny ship to sail in her wake. She is almost ready to be launched as I write.

A surviving seventeenth-century chart shows where the Dutch sailors first saw the Australian continent, at Cape York Peninsula, and the Aboriginal elders of Cape Keerweer told me their traditional story of the first fatal encounters with the Dutch: 'We defeated them ... and we can show you where our people buried some'.

In 1581, when the King of Spain's control over Portugal excluded the aggressively entrepreneurial Dutch from the port of Lisbon, with its lucrative East Indies trade, the Dutch began thinking of sending their own ships to the Indies.

In that year Prince Willem I of Orange, the Stadholder (leader) of Holland, Zeeland and Utrecht, and the States-General of the Seven Provinces, formally renounced their allegiance to Philip II of Spain. Willem opposed the Spanish Inquisition, fought valiantly for the freedom of the United Seventeen Provinces and had been by turns a Romanist, Lutheran and Calvinist, visualising a state in which Protestants and Roman Catholics could live on terms of mutual respect and equality, or at least toleration.

In July of 1584 Prince Willem was assassinated at his home in the city of Delft by a young fanatic from Burgundy named Balthasar Gerard, who was immediately branded as an agent of the Spanish crown. There was thus no love lost between the emerging Dutch republic and the despots of the Iberian peninsula, an enmity that would soon manifest itself in the struggle for the East Indies.

After Christopher Columbus returned from discovering what he thought was India, the Spanish Pope Alexander VI issued a decree granting Spain all lands discovered, or yet to be discovered, in the regions explored by Columbus. He drew an imaginary north–

south line 100 leagues (480 kilometres) west of the Azores and Cape Verde Islands, and proclaimed that the land and sea beyond that line should be a Spanish sphere of exploration. After some negotiations the Portuguese monarch agreed on the basis that the Pope's line be moved 270 leagues (1300 kilometres) further westward.[1]

This arrangement of 1494, which became known as the Treaty of Tordesillas, awarded Portugal most of the southern Atlantic and Brazil, as well as the eastward trade route to India. There would be much argument between Spain and Portugal in the ensuing century about where exactly the Pope's line ran down the other side of the world in the region of the Philippines and the Maluku Spice Islands. For their part, the late sixteenth-century Dutch merchants were not in the mood to respect the aspirations of their Spanish oppressors, or the encyclicals of the Catholic Pope, who symbolised a generation of vicious religious persecution in their vigorous, emerging homeland.

Initially the Dutch, unlike their Iberian predecessors, were reluctant to possess and control new territories, but when they discovered how plentiful and cheap spices and other sought-after products were throughout the East Indies they set out deliberately to create a monopoly for themselves, by force of arms if necessary, over the European market through control of the supply lines from the East Indies.[2] Later they would create a landed empire in South-east Asia, that would last well into the twentieth century.

Spices and peppers were highly valued for making stale meat more palatable in northern Europe after the annual slaughter of livestock before winter, and oil of cloves was used as a medicine for various ailments and even claimed to enhance the enjoyment of lovemaking. The oldest remains of cloves are said to have been found in Syria's Mesopotamian tombs dating from about 1800BC.

Indonesia today imports cloves, but in the sixteenth century, when the first Portuguese and Spanish explorers found their way to the East Indies the world's entire supply of cloves came from the Maluku islands of Ternate, Tidore, Motir, Makian and Bacan, off the western coast of the big four-armed island of Halmahera. Cloves were an essential part of the traditional mythology of the world of Maluku. Much of the pepper brought to Europe in the fifteenth century came from the western coast of India, but the best was grown in Sumatra. Cinnamon came from Ceylon, nutmegs and their by-product mace chiefly from the Banda Islands.[3]

There was an insatiable demand in the near east and all over Europe for cloves. For many centuries, far eastern cargoes for Europe had travelled overland on camels and donkeys by many different caravan routes, to terminate in the eastern Mediterranean ports of Turkey, Syria, Lebanon and Palestine, and on the Black Sea. Chinese traders collected the cloves and nutmeg of the East Indies in their sailing vessels, and delivered them for sale in the great south-west Malayan port of Malacca. From there across the Bay of Bengal to India the trade was in the hands of Muslim merchants. (Islam had reached the East Indies during the thirteenth century.) In India the spice cargoes were sold in the south-west ports of the Malabar Coast and further north. The trade from there was largely in the hands of Arabs.

As the cost of spices and peppers soared in the late fifteenth century, western Europe began to look seriously for ways to break the Arab monopoly. An all-sea route to India was becoming more of a possibility as Portuguese explorers sailed farther down the Atlantic. Portuguese navigator Bartholomew Diaz rounded the southern tip of Africa in 1488 and another Portuguese, Vasco da Gama (c.1469–1524), sailed past there in 1497 and discovered a sea route to the western coast of India.

Encouraged by Prince Henry the Navigator, Portuguese explorers continued eastward in the wake of Vasco da Gama. Admiral Alfonso de Albuquerque, sent out with a plan to break the Muslim domination of vital centres along the route, captured Goa on the western coast of India in 1510, and in the following year seized Malacca.

Several months later de Albuquerque sent three of his ships, commanded by Antonio de Abreau, to search for the Maluku Spice Islands.[4] They sailed eastward past Java to the islands of Buru, Amboyna (or Ambon) and Seram, where Captain Francisco Serrao ordered his old and unseaworthy ship to be burnt before they turned SSW to Banda. Here Serrao bought an Asian junk and loaded it up with cloves, nutmeg and mace.

But on the return voyage from Banda the fleet ran into a storm in which the junk ran on a shoal and was sunk. The survivors, only Serrao and half-a-dozen of his crew, were rescued by some local fishermen and taken to Amboyna. Their presence was reported to the Sultan of Ternate (the main cloves island, north of Seram), who was eager to gain an ally in the powerful Portuguese nation about which he had heard much from traders and travellers. The sultan sent his brother with a junk to welcome Serrao and his men to Ternate, where control of the most precious spice island was then virtually handed to the King of Portugal.

In 1519 Ferdinand Magellan (c.1480–1521), a Portuguese navigator in the service of Spain, found the strait leading around South America to the Pacific and sailed across to the Philippines, where he was killed in a pointless involvement in a local battle. But one of his two ships which continued to the Maluku Spice Islands, picked up a small cargo of spices and in 1522 completed the first circumnavigation of the world. This was so profitable as to convince the Spanish royal court it should sponsor further voyages.

The Sultan of Tidore, traditional enemy of Ternate, took advantage of the arrival of two remaining ships of Magellan's expedition in November 1521 to seek a similar arrangement with Spain. The Spaniards promptly accepted Tidore, but their several expeditions which followed were failures, and in 1529 Charles V of Spain, at war with France and almost insolvent, signed the Treaty of Saragossa, pledging all rights in Maluku to Portugal for 350,000 ducats. An arbitrary line of demarcation was fixed seventeen degrees east of the islands (about 1900 kilometres) and Tidore was handed over to the Portuguese.[5]

In 1528–29 the Spaniard Alvaro de Saavedra, commanding the ship *Santiago*, sailed from New Spain (South America) for the East Indies, and returning eastward on a more southerly course along the northern coast of New Guinea, named it Isla del Oro ('island of gold'). About this time, old charts show, the western end of the north coast of the island was known as Os Papuas.

Another Spaniard, Inigo Ortez de Retez, commanding the *San Juan*, would rename de Saavedra's 'island of gold' in 1545 as Nueva Guinea. It was said that this name came from a description of the local people, similar to the dark-skinned woolly-haired people of the African Gulf of Guinea.[6]

In 1521 the Portuguese had sent a governor to Ternate and he erected a fort beside its royal town of Gammalama. The Portuguese masters of the island now behaved abominably toward the Ternateans, with a mixture of aggressive Roman Catholic proselytising, rape, murder (especially by poisoning) and countless intrigues designed to escalate the conflict between Ternate and its near neighbour Tidore.[7]

The energetic Francis Xavier came to Ternate in 1547 in search of souls, and while there he enjoyed a friendly relationship with the Maluku Sultan Hairun, but this did not prevent the Portuguese from imprisoning Hairun for refusing to give his lands to them. While in prison he escaped a Portuguese plan to poison him and, disenchanted with the Roman Catholicism he had experienced, became a strong defender of the Islamic faith. But eventually on 28 February 1570 the sultan was assassinated on orders of the new Portuguese Captain Diogo Lopes de Mezquita.

Outraged by the murder, the Ternateans, led by the new sultan, Hairun's son Baabullah, laid siege to the Portuguese Fort Gammalama. Five years later the Portuguese were forced to make a hasty exit from Ternate. Baabullah ruled from 1570 to 1583 in a golden age of prosperity and expansion while his warriors harassed the Spaniards in the Southern Philippines and strengthened ties with the Muslim communities of Mindanao. Resistance to the Portuguese began to spread throughout Maluku.

When Sir Francis Drake called at Ternate in 1579 during the first English circumnavigation of the globe, Sultan Baabullah welcomed him with some splendour, dealt him a good cargo of spices and offered his allegiance to Queen Elizabeth. Portugal and Spain were united in 1580, and four years later they combined to seize the island of Motir and hand it to the sultan of nearby Tidore, but they failed to capture Ternate. Baabullah was succeeded in 1584 by his son Sultan Said al-din Berkat Syah.[8]

IN THE Netherlands in 1595 preparations were completed for the first Dutch expedition to the East Indies. In May of the previous year the keels had been laid at Uylenburg, near Amsterdam, for the four ships: *Hollandia* and *Mauritius*, both of 460 tons), *Amsterdam* (260 tons) and *Duyfken*, of only fifty or sixty tons.

The popular military leader Prince Maurits had agreed to christen the new ship *Mauritius* during the victory tour for his liberation of the northern town of Groningen from the hated Spanish control. This was a great honour for the Amsterdam merchants involved, and a matter of pride for the young nation. Near the end of February 1595 all four ships had been launched, rigged and loaded, and were assembled in busy Amsterdam harbour. Soon the four would move up the Zuiderzee, close to the entrance to the North Sea, to take on the last of the crew and provisions and await favourable winds.

This first *Duyfken*, as shown by engravings of the time, was built like the other three vessels in the fleet, with three masts, square-rigged on the fore and mainmast, but only about twenty metres from stem to stern, with a draught of about 2.3 metres. She could sail anywhere with the rest, but was also suitable for carrying messages or people between ships, for venturing closer to shore and exploring in unknown waters. She carried ten cannon and other arms — more like an armed carrier pigeon than a traditional dove of peace.

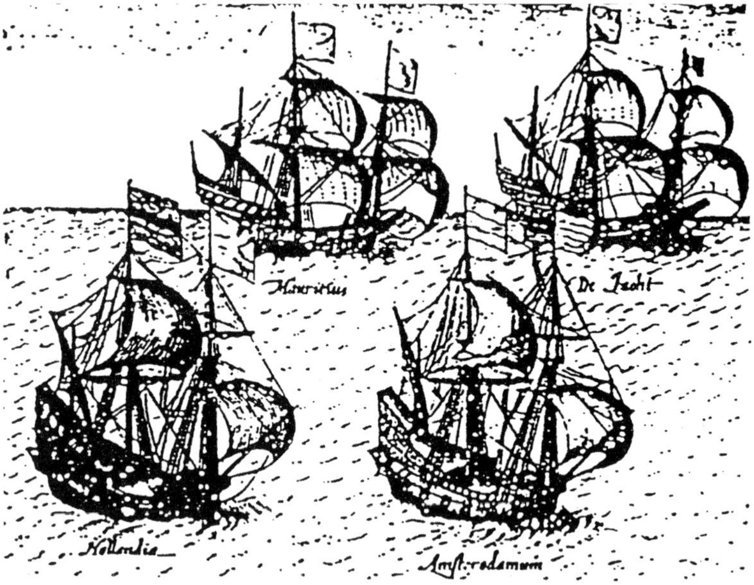

The four ships of the Dutch first fleet of 1595, from a seventeenth-century engraving: *Mauritius*, *de Jacht* (*Duyfken*), *Hollandia* and *Amsterdam*.

On 2 April 1595 the four ships set off from the northern Island of Texel, crept down the Atlantic at speeds of three or four knots, almost to the Abrolhos Islands near Brazil, then continued south until they could pick up the winds to take them eastward to head up the East African coast, following the well-established but laborious route of the pioneering Portuguese. By September, when they reached the island of Madagascar, one skipper and about seventy crew were already dead from scurvy and many of the survivors were sick from lack of fresh food. Some were sent ashore to die. There was a long delay and a rearrangement of ships' commanders and crews, causing many problems and protests; with about 6500 kilometres yet to go to the Indies.

On 6 June 1596 the fleet arrived at Engano, a small island off south-western Sumatra. They obtained water and food and stayed seventeen days before catching suitable winds to take them eastward and through Sunda Strait to the little city of Bantam (or Banten), on the north Java coast.

The Dutch found that Portuguese traders at Bantam, supported by the sabander (a senior official like a harbourmaster controlling customs and excise), the governor and probably a local sultan, were not prepared to allow free Javanese trade with the Dutch. Some Dutchmen who went ashore, including the fleet commander Cornelis de Houtman, were seized as hostages.

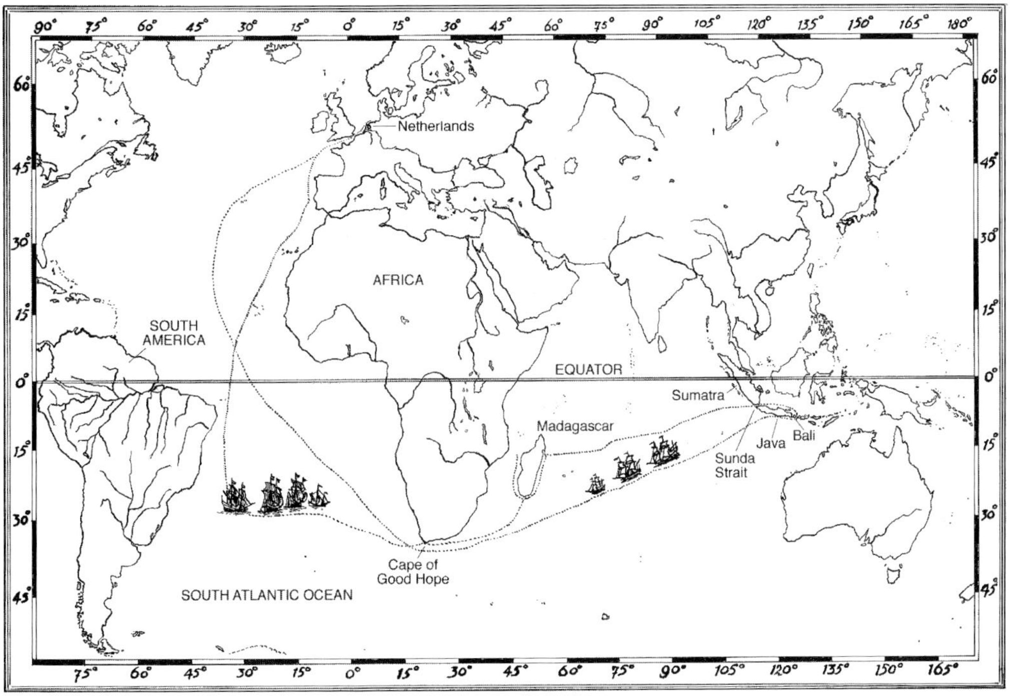

The outward and return journeys of the first Dutch fleet to the East Indies, 1595–97.

The fleet reacted by delivering an ultimatum: to release the hostages and sign written statements agreeing to free trade — or else. Messages were received from the Dutch hostages ashore, that the sabander had promised to release them and negotiate trading terms. However, while the fleet was waiting for the sabander to honour his words, he hastily built up Bantam's shore defences with Portuguese artillery.

On 7 September the officers commanding the Dutch fleet decided they had waited long enough. What happened next is depicted graphically in seventeenth century accounts and engravings which indicate that the *Duyfken* and her crew were the heroes of the day.

The three bigger Dutch ships sailed as close as possible to the city of Bantam and pounded the shore with their heavy guns while the *Duyfken* cruised in closer to blast the city with her cannon, particularly two lethal short-barrelled heavy guns. These inflicted most of the fleet's damage and killed about fourteen Bantam residents while the Portuguese vainly fired back about thirty cannon shots from the city walls.

Amid the cacophony of battle a Javanese junk sailed serenely out of the harbour. The Dutch commander signalled the *Duyfken*'s skipper to pursue the potentially rich prize, but it proved to be a trap, luring the *Duyfken* into shallow water where, despite her light draught, she ran aground.

At this point, a fleet of twenty-four Javanese prahus (or praos), each carrying sixty warriors, sped out and advanced in a half-moon battle formation approaching *Duyfken*. But just before they could reach her, the Dutch used a ship's boat to carry an anchor out to deeper water and hauled *Duyfken* off the bank, using the ship's windlass.

Sailors hastily cut *Duyfken*'s anchor cable as the enemy boats converged on the ship's boat. By now the dauntless vessel was pitted against a swarm of about 1000 eager Javanese warriors brandishing spears, raising bows and arrows and preparing to board, but the Dutch skipper coolly held fire until the Javanese were 'looking into the muzzles of the cannon'.

Then from each side of the little ship five cannon flamed and roared, smashing iron balls into the tightly packed ranks of Javanese and slowing their prahus as the ship gradually got under way and escaped the trap.

Duyfken's cannon then flung heavy shot on to the enemy formation, sinking several prahus, but some brave warriors swam from one as it sank, and clambered into the trailing ship's boat, cutting the towing line. Then they furiously rowed their prahus alongside to jab their spears into the gunports. However the interruption was brief. *Duyfken*'s gunners repelled these attackers with a fusillade of musket shots and the ship made good her narrow escape. Sailing back to join the other three ships, she continued to rake the prahus with shot, killing more than 100 warriors before rejoining the attack on Bantam.

More messages were received from the Dutch hostages, now urgently appealing for cessation of Dutch fire because the sabander had threatened 'to put them on stakes'. The Dutch fleet had no alternative but to cease operations to await further developments, and by 13 September the lack of drinking water and provisions forced the fleet to leave Bantam and their people held hostage ashore, returning on 1 October. The following morning a letter was delivered to the Dutch fleet, demanding a payment of 2000 reals of eight as damages and ransom to free the Dutch hostages.

The fleet command decided to pay up, and from then on, local leaders were allowed to sell provisions to the Dutch. On 22 October Admiral Cornelis de Houtman and the other hostages were safely back on board the ships. But the trade levies demanded by the sabander (apparently on advice of the Portuguese governor) were excessive and although some minor trading took place the Dutch fleet decided to sail on eastward and try their luck elsewhere. Fortune continued to elude them, however.

On 5 December 1596, at an anchorage by an island off the town of Sedayu at the other end of Java, pirates boarded the *Amsterdam*, and their pirate 'prince' murdered the skipper Jan Jacobs Schelinger by stabbing him with a Javanese dagger. They were repelled after killing twenty-seven crew.

Ten days later, a monsoon swept down from the north-west and prevented the ships from moving from shelter for eleven days. When Commander de Houtman ordered them to continue the expedition to the northern Spice Islands there was a mutiny, and the council of ships' officers decided it was time to go home.

The fleet paused for several weeks at 'Paradise Island' (Bali) for the demoralised crews to take on water and fresh food for the difficult return voyage. By now, not enough crew remained alive to sail the badly damaged *Amsterdam* home, so they burned and abandoned her by the island of Bawean, off the north coast of Java, north of Surabaya, on 12 January 1597.

After seven more months of tribulations, the bedraggled fleet arrived at the Dutch port of Petten, almost home, on 10 August 1597. Next day *Duyfken* and *Mauritius* weighed anchor and sailed the remaining fifteen miles to a dismal welcome at Den Helder and Texel Island. *Hollandia* could not join them until four days later because her exhausted crew had been unable to raise their anchor.

The first expedition was a lamentable disaster. Only one of the four skippers, Simon Lambrechtsz Mau, the original skipper of the *Duyfken*, survived the ordeal; only eighty-nine of the total 249 crew returned alive and seven were carried ashore to die soon afterwards. The venture made little profit, but the Amsterdam merchants were not deterred, having learnt more by now from other sources, such as the travel writer Jan Huygen van Linschoten, about the rich potential of the islands of spice and pepper. At the very least, the de Houtman expedition proved that it could be done.

The next, better organised, Dutch expedition to the East Indies, departed on 1 May 1598. Six ships were named after the provinces of Holland: Zeeland, Utrecht, Gelderland, Friesland and Overijssel. The seventh won the name *Mauritius* (for the Stadholder) and the eighth was named for the city of Amsterdam. On *Hollandia*, one Willem Jansz served as first mate. This second venture to the East Indies, and many more which followed, made great fortunes for shareholders.[9]

Section of the banner reproduced in full on pages 62–63. Details from the engraving — of the busy Dutch port of Amsterdam — are placed throughout the book.

The Linschoten Society volumes[10] reveal much relevant information about the first two Dutch expeditions to the Indies. An English language section was headed:

> A TRUE REPORT of the gainful, prosperous and Speedy voiage to Java in the East Indies performed by a fleet of eight Ships of Amsterdam WHICH SET FORTH FROM Texel in Holland, the first of Maie 1598 Stilo Novo [new style]. Whereof foure returned again the 15 of July Anno Domini 1599, in less than 15 months, the other foure went forward from Java for the Moluccas.

A footnote explained Stilo Novo as the change in calendar from the Julian to the Pope Gregory XIII as from 24 February 1582.

The report referred briefly to the first voyage of 1595 and then turned to 'the fleet of eight gallant ships' of the 1598 expedition which sailed 'in hope of better successe'. The list of ships along with their skippers and merchants ended with:

> … a pinnas that had been in the former voyage, called *The Pigeon*, now y [yacht?] *Overijssel*, of the burden of fifty tuns. The master Symon Johnson [or Jansz].

Another sixteenth-century document, *Historical Description of the Widely Renowned Merchant Town of Amsterdam*, printed in 1614 by the Dutch cartographer Jodocus Hondius, confirmed that the *Overijssel* was the 1594 *Duyfken* under a new name. It contained an account of the 1598 expedition to the Indies and: '*Overijssel* was lesser than the others, and abler in sailing, and was also called by another name, the Dove'.[11] This *Duyfken*, for many years thought to be the same vessel that Willem Jansz unwittingly sailed into history in 1606, is thus revealed as only the first in a long line of ships to bear that name.

The Next *Duyfken*

ON 23 APRIL 1601 another small ship named *Duyfken* was one of five vessels of a Dutch fleet which put to sea from the roads of Texel. She was described in the shipping records as a sixty-ton yacht, and this would be her first voyage to the Indies.

In command of the fleet, equipped by the Oude Oost-Indische Compagnie (Old East Indies Company) of Amsterdam, was Admiral Wolphert Harmensz, on his flagship *Gelderland* (360 tons). The others were *Zeelandia* (360), *Utrecht* (240), *Wachter* (130) and the yacht *Duyfken* (60). The fleet was to sail via South Africa's Cape of Good Hope to the Maluku and Banda islands, take on a full cargo of spices and return to the Netherlands.

The two journals kept aboard *Gelderland* are preserved by the Dutch National Archives General,[12] and both carry references to *Duyfken*. Almost half of the first volume consists of diverse drawings of fishes and birds — including the dodo, the flightless bird of Mauritius which was wiped out by man in the seventeenth century — and panoramic views of coastlines. The draftsman in this first journal was clearly not the author of the text: the inscriptions for the drawings were in a different hand from that of the main text. Who the authors were could not be determined from the journals; nowhere did they mention their names.

The other journal consisted of drawings of pottery, some fishes and a short series of panoramic maps called 'afconterfeytinghe' — which translated directly from the old usage as 'counterfeits'. The maker of these pictures was justly proud enough of his work to inscribe it with: 'by me, Jooris Joostensz. Fecit'. ('Fecit' was a Latin suffix, 'has made'.) In three of these 'afconterfeytinghe' Jooris Joostensz made sketches of the ships in the fleet. It was possible to positively identify the yacht *Duyfken* in one of the drawings, because the *Gelderland* journal stated that when she left Banda in the East Indies for Bantam, she was in company with *Zeelandia* and *Duyfken*, while the other two ships stayed behind to wait for more cargo. Joostensz showed the three ships at anchor near Tuban, on Java, and in this sketch the tiny *Duyfken* could not be mistaken. The ship was shown here with her topsail yards sent down.

In the other two drawings of the fleet of five riding at anchor there was again only one small ship with no top yards crossed. This feature, and her relative size, distinguished *Duyfken* from the others. The small ship was drawn without a gallery at the stern in two of the sketches, and also in the third, though the latter is such a small sketch that the gallery might have been simply omitted from it. Finally, the appearance of this ship as a sleeker vessel than the others concluded a positive identification of *Duyfken*, the ship which would sail to Australia.

The inscriptions Jooris Joostensz included in his drawings were of a scale of Dutch 'mijlen'. There were fifteen Dutch mijlen to one degree of sixty nautical miles: one mijl equals about four nautical miles or just over seven kilometres. Latitude and sometimes longitude of places were given. There were names of towns and settlements visited on the voyage, references to other features such as special mountains or reefs; locations where the fleet obtained water, fresh food or where a camp was set up for the sick. Seven of the eleven maps contain a compass rose, making a combination of a mariner's guide and a portolano chart (of directions, descriptions of harbours and so on).

Under favourable conditions the fleet could maintain about five knots for prolonged periods. And the *Duyfken*, although the smallest ship in the fleet, actually out-performed the others, which according to an entry of 15 September 1601, could not catch up with her despite a fresh breeze and all their topsails set. In unfamiliar waters the fleet moved cautiously, especially when diminishing depths were sounded. *Duyfken*, with her shallow draught and manoeuvrability, was shown to have an active role to play in assisting the fleet in these situations. The main journal opened:

Journal, or daily register, beginning at 22 April A. [Anno Domini] 1601 in the ship *Gelderland*. Ditto the lord admiral has come aboard from Huysduijnen.

23 Ditto went under sail in the morning the wind North East.

The voyage seemed initially uneventful, except for a brief entry in June:

2 [June] Ditto Wind still. Today someone was keeled under the yacht for thievery. [The punishment of keel-hauling was likely to be fatal.]

The voyage continued southward to 35°30′ south latitude by the middle of August 1601, then followed an easterly course until the beginning of September to pass the Cape of Good Hope, and turned north-east toward the island they called Mauritius (in honour of Prince Maurits). Until the end of September there were serious problems in finding the island, but the final anchorage was reached early in the morning of 1 October. Even then, Admiral Harmensz was doubtful:

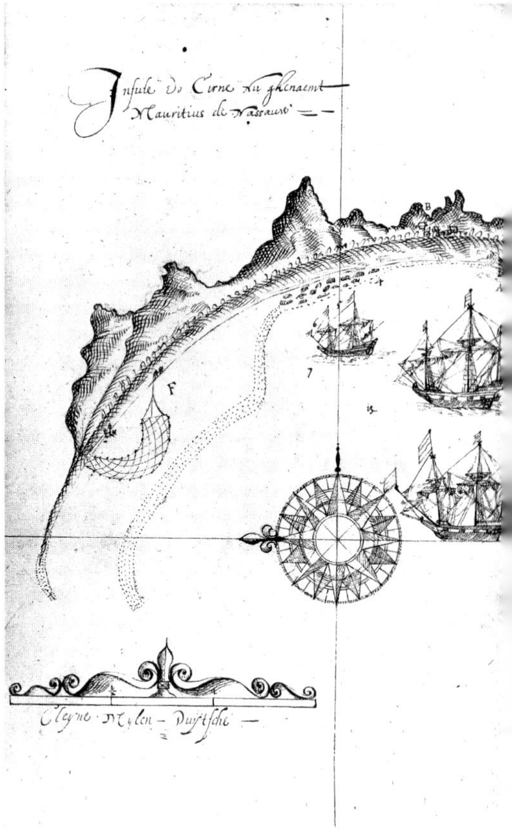

… and in order for us to know whether the island is Mauritius or not, so has the admiral ordered the yacht [*Duyfken*] to the harbour to see … On 4 October the yacht has left us to sail to the harbour & has arrived [there] on the 7 ditto in the evening.

On the 8 ditto they went ashore & found there a naked man who was a Frenchman who had been there on his own for 20 months & had sustained himself with palmites, cocos [coconuts] and dried fish, because he has been without fire. He had gotten there in a Jonck from the islands of East Judijen [??], because that's where he went with an Englishman & [they] lost their ship there according to what he said.

The 12 ditto the yacht [*Duyfken*] has tacked out of the harbour & coming outside in the evening has set her course toward us & have come back with us on the 13 ditto & brought us confirmation that it was Mauritius, and [they] have brought the naked man along & so we got ourselves water & firewood & readied our ships to sail for Java.

This said something about *Duyfken*'s windward ability. There were references scattered throughout the *Gelderland* journal, to the windward ability of the fleet, and these suggested the ships could stand as close as twenty-five or thirty degrees to the wind. It was impossible however to determine if they were also making any distance good to windward.

The Frenchman, whom they called François, stayed on board the *Gelderland*. But his story was to end sadly; he would die on 25 May 1602 of 'natural causes' and be buried on the island of Banda, in the East Indies.

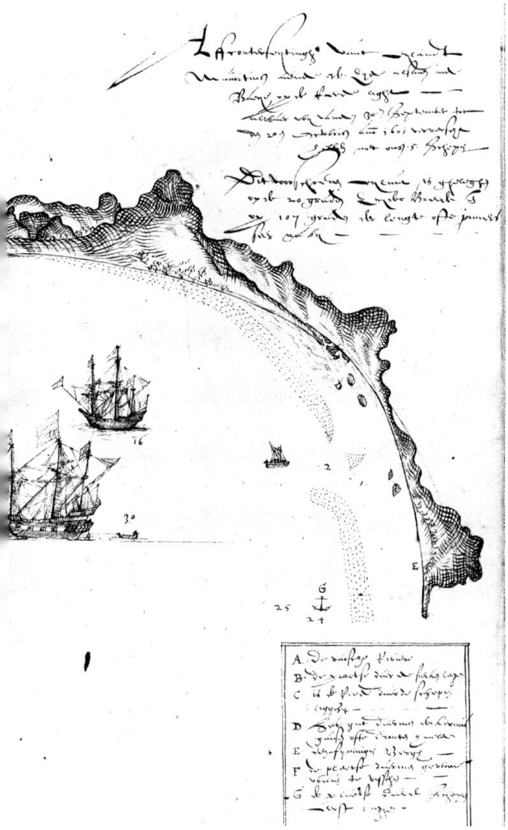

The *Gelderland* fleet at the island of Mauritius in October 1601, *Duyfken* identified at top left. (From the *Gelderland* journal, courtesy Rijksmuseum, Amsterdam.)

On 13 October 1601 there was a meeting of the Broad Council of senior officers of the fleet in which some interesting decisions were made:

At present, 13 October 1601 being under the island Maurits at the roadstead not unjustly called 'riggers' [roads], the College of the Broad Council was called into meeting by the lord admiral to make some necessary changes regarding some matters.

[It] is firstly resolved by the College and ordered as proper that Abram Francken, understeersman, because the yacht *Duyfken* is undermanned with her 3rd watch, shall transfer from the ship *Utrecht* to the yacht [*Duyfken*] as uppersteersman [navigator] enjoying improvements about this to the discretion of the directors of this Comp. residing in Amsterdam — thus acted in the ship *Gelderland*, the 13 October 1601.

Secondly because the vice-admiral and his skipper Jan Cornelisz [of the 360-tonne ship *Zeelandia*] have uttered complaints as being badly served by a certain understeersman tending to be utterly useless and moreover does not know about sailing or anything pertaining to ship matters, being less than least in the fleet, nor having made any effort ever to learn about it, so that in the common interest the upper-bosun will take over the position of Jooris Joosten, understeersman, and Jooris Joosten shall transfer to the ship *Gelderland* and serve in whatever function the Admiral shall give him; because of this aforem. Symon Willems will enjoy improvements to the discretion of the Comp. — thus acted in the ship *Gelderland* 13 October A.1601.

So now it could be seen how the unwanted Jooris Joostensz, whose preserved sketches showed him as a skilful artist, ended up on board the *Gelderland*. In 1997 Joostensz' section of the *Gelderland* journal had an insured value of $800,000 as one of the prize items of the 1997 'Hartog to de Vlamingh' exhibition at the Western Australian Maritime Museum. Abraham Francken, promoted to uppersteersman of *Duyfken*, would enter the *Duyfken* story later.

The fleet departed from Mauritius on 20 October 1601. The journal of 28 November included a bizarre shipboard tale of a demon being exorcised from an occasional drunkard, a young sailor named Jan Lambertsz who had been noisily possessed by the devil. The spellbinding exorcist was named as 'Melisz Joorisz, the comforter of the sick'.

By 26 December 1601, the fleet had crossed the Indian Ocean and sailed into the Sunda Strait almost to Bantam. Near a little town called Palimba a Chinese prahu came alongside 'to bring us tidings ...' which amounted to the warning of an ambush just around the corner:

> ... that before the town of Bantam lay a Portuguese armada of 30 sail, as he said, so many ships, galleys, fusten [similar to galleys] and frigates that it made a great fright amongst us.

The Dutch fleet here comprised only five vessels, so Admiral Harmensz ordered a meeting of the Broad Council to deliberate on the next course of action while his fleet anchored off the town of Palimba for the day. Despite the odds, the officers voted with a unanimous show of hands:

> ... that we should ready ourselves by the next day and be well provided with everything in order to fight our enemy boldly and united in arms to make our name reckoned within that country.

Early next morning (27 December 1601) the small Dutch fleet raised anchor and sailed around Tanjung Pojok, at the northern end of Sunda Strait. Sure enough, there ahead were just as many sail as the Chinese had warned, to the east and in the Bay of Bantam below. The armada included eight large ships, two of which were on watch.

The Dutch sailed boldly on, their cannon inflicting much damage on the enemy, until an explosion burst open one of the cannon in *Gelderland*'s stern battery:

... the whole of our rudder [was] damaged because the pintle of the rudder was [so] completely shattered that we could accomplish nothing more, then the other ships came to help the other two [enemy].

Then we had a rain squall and wind and current shifted that we could not get at the Portuguese ships and the wind came from the W.S.W., so we dropped our anchors to await a shift in the wind such that we could do another sortie against her, and continuously we were in prayers, entrusting God that He would provide us with victory over our enemies.

The unfavourable wind and current lasted until 29 December. Then:

... and the wind was North West that we could sail a course toward two of their fusten [Javanese boats] which had broken anchors the day before, so we went under sail again and have taken them, of which those of the crews putting up resistance were slain and the others taken in mercy, and [we] have set one fust on fire an left the other adrift.

The Portuguese ships were not brave enough to close up on us because they lay about 1.5 mijlen [eleven kilometres] to windward from us, and we set there and they did not dare come to pelt us with shots so we lay there until the 30 ditto.

In the morning (of 30 December) we saw a fusta that lay toward the land from us and Jan Martens' boat [the ship's boat of *Utrecht*, possibly skippered by Jan Martensz] with 12 men and 4 muskets. That fust, seeing this, cut his anchor and sailed near the shore and then at least 20 or 25 men jumped overboard to swim ashore, and still 30 or 40 men stayed on board and kept it going in the breakers that our folk did not dare join her and so came back again.

The Frenchman found at Mauritius. (From a drawing in the *Gelderland* journal, 1601.)

Further this day they [the Portuguese] have also set three fusten on fire and put their crews on board the large ships for reinforcements, and because of this the fustas were easy to be taken.

A meeting of the Broad Council resolved to send a pinnace during the night with a letter to the King of Bantam to advise him what had happened. But the mission was unsuccessful; the pinnace returned about midnight because the current was too strong for them.

Next morning, 31 December, the Dutch fleet put up sail to get to windward of the armada, but the enemy did the same, and the Dutch 'lost much ground'. Then the Portuguese came forward to scare them. Observing this advance, the Dutch took in their sails and dropped anchor. So the Portuguese also dropped anchor, being afraid now to come within range of the lethal Dutch cannon.

The decisive Battle of Bantam between the Dutch and Portuguese. *Duyfken* is in the central foreground, firing at a Portuguese galley. (Engraving by C. J. Visscher and H. Allarts, courtesy Rijksmuseum.)

In the morning of 1 January 1602 the Harmensz fleet raised sail again and got to windward of the armada. They shot at the closest enemy ship but the Portuguese then fled from the town of Bantam and passed behind the wake of the Dutch ships.

The Dutch anchored again, and on the second day of January all of their ships headed back in for Bantam where they dropped anchor about 7 a.m.

... and we were very cheerful and had a very festive reception by the burghers and officials and stayed there until 12 ditto.

A letter written aboard *Duyfken*, unsigned but evidently by an officer with detailed knowledge of the Harmensz fleet, related that the Chinese who brought the warning of the Portuguese armada had come aboard *Duyfken* from his prahu and that the little ship had sped to the flagship with the message. The *Duyfken* letter gave a dramatic account of what would become known as the Battle of Bantam.[13]

More luridly detailed was the description in a booklet produced about seven years later by a maritime author of the time named Cornelis Claesz,[14] who wrote that during the battle with the armada off Bantam the *Duyfken* came to the assistance of a Dutch ship that was being boarded by the crew of an enemy galley. Most of the boarders then jumped overboard. Of the Portuguese captain and his twenty-three crew along with sixty 'blacks or Indians': The captain and some Portuguese were taken on board. The rest were mostly beaten to death.

From another captured enemy galley the elderly Portuguese captain did not want to give himself up and be taken prisoner, Cornelis Claesz wrote, 'So he was run through with a pike'. And later in the description, there was another casual reference to the fate of several more captains being taken prisoner and 'beaten to death by [the Dutch]'.

The writer of this *Gelderland* journal noted that the Portuguese armada was under the command of Admiral Andre Furtado de Mendoca. The Dutch would encounter him elsewhere later.

April 1602 found the Harmensz fleet anchored, loading spices at the south Maluku island of Banda, below the then partly explored island of Ceram — it runs about 330 kilometres from west to east and is spelt Seram on modern charts, with Seram Sea to the north.

On 10 April the admiral called a meeting of the Broad Council, which resolved to send *Duyfken* on a reconnaissance-trade voyage north to the large island of Ceram and then east to some other islands. The council drew up the following instructions for the upper-merchant Claes Gaef and skipper Willem Cornelisz:

First he shall sail to the island of Cenam and these following harbours and roadsteads such as Quevin, Quelibara, Quelilonhen or Gonlegonlij at fault of these [or instead], to some others where one may expect good trade. No longer than fifty days shall you stay there after arrival, but the shorter the better if you can obtain your cargo sooner. And if it happened what God forbid, that in the prescribed time the yacht were not laden, you will in any case be remunerated here by us in Banda for what you may have obtained.

Secondly, if something more than sago can be obtained there, their ways of trading to what places what wares are brought [that are] of use there and to what limits her farthest shipping is stretching. Also if they have knowledge of Nova Ginnea, and if ever ships have sailed from there or have come from Nova Ginnea to Cenam.

At the island of Banda thus acted the 10 April A.1602 in the ship *Gelderland*, God send His Blessings unto salvation. Amen.

Duyfken returned within thirty-five of the fifty days allowed for the task, and on 15 May 1602 the *Gelderland* journal reported from the *Duyfken*'s account of 'certain islands with which they of the islands of Cenam and Banda carry on trade'. Apart from Cenam (or nearby Ceram or both), it stated that *Duyfken* called on islands named Papou, Lille and Tymaer.[15]

Apparently *Duyfken* had visited the island of Misool, south of the western end of New Guinea where, the journal reported, 'there is great demand for cloth, gold, gongs, iron and swords'. Dealing with the 'Nova Ginnea' inquiries, it was summarised in the margin of *Gelderland*: 'Note what they of Ceram know about Nova Guinea is very little'. And the main entry raised an interesting question:

About that island of Nova Ginnea, no knowledge except for saying that white people[16] live there on the south side. [If it was] inhabited by Portuguese yet the local people had yet never seen Portuguese ships. They cannot say anything about their trade nor merchandise.

The *Gelderland* journal on 23 June 1602 (after *Duyfken*'s exploration east of Banda) stated that the Dutch ships were still at Banda, and next day *Gelderland*, *Zeelandia* and *Duyfken* left for Bantam, arriving there on 1 August, but *Wachter* and *Utrecht* stayed on at Banda to wait for more cargo.

At Bantam it was decided on 15 August that *Duyfken*'s uppermerchant Claes Gaef and supercargo (chief merchant officer) Jan Lodewycksz Roossengijn would stay in Bantam to await the arrival of the ships *Utrecht* and *Wachter*. The flagship and its two accompanying vessels departed for home on 25 August 1602.

On the return voyage, one night *Gelderland*'s sails were reported blown to pieces. Separated from the others (not unusual in long voyages) she found her way to the island of St Helena, about 1850 kilometres west of Angola. *Zeelandia* arrived there next day, and the *Gelderland* journal of 25 November 1602 noted:

... In the morning our vice-admiral [the ship *Zeelandia*] arrived in the roads thinking that the yacht [*Duyfken*] was also coming, but they said it had separated from them at the Cape [of Good Hope] because it had been sent to look for us.

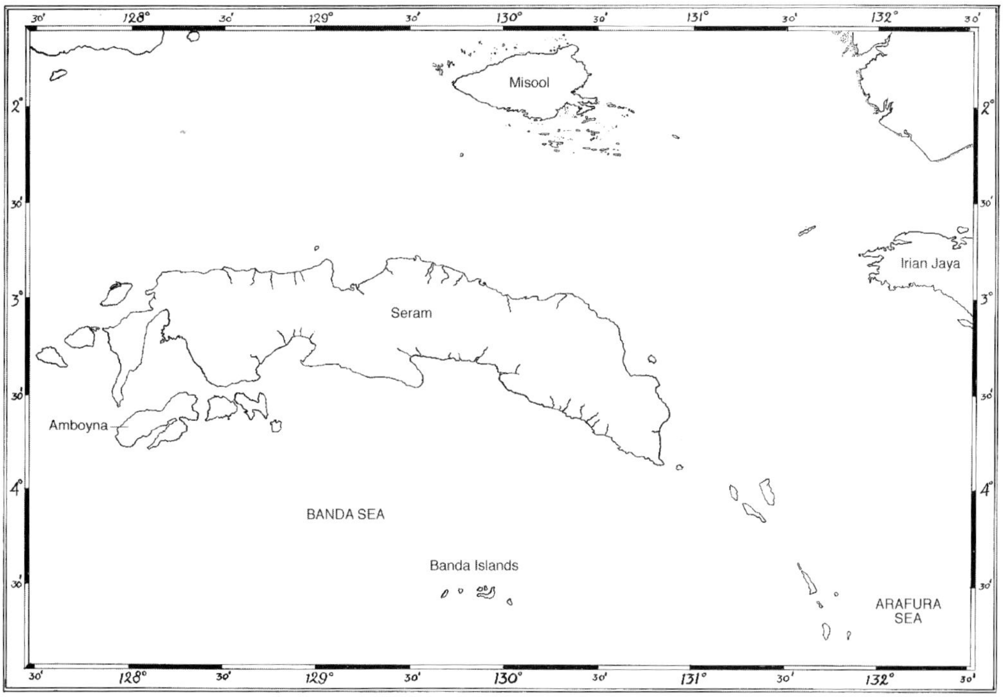

Seram and its neighbourhood.

They waited more than three weeks, and it seemed likely *Duyfken* had been lost. The two ships left on 22 December and several months passed before the last three entries would be made in the daily register of *Gelderland*:

24 March 1603 in the morning spoken with a small ship which came from Rowanen [Rouen, France] and understood from her the situation of the country which caused great joy among us. [This may have been encouraging news from home of victories in the war against Spain.]

In the evening we got the wind from the north-east with stiff continuous breezes. Laid our course over Northward to England.

25 ditto in the morning we bore away straight for Ramshooft [Rame Head, near Plymouth]. In the evening [because of the NE wind?] we came into the roads of Pleijmuijen [Plymouth] and understood there that our yacht [*Duyfken*] had safely arrived in Middelburgh [in the Dutch province of Zeeland], which was cause for much joy to us. The Lord be praised and thanked for His great Love and Mercy to us ever, amen.

On 27 March *Gelderland* and *Zeelandia* arrived at 'the West Confer' near Plymouth and they stayed there until 14 April 1603.

The *Duyfken* letter mentioned above placed on record her safe, early arrival at Vlissingen (Flushing) on 17 February 1603 'hoping that the other ships will have got to St Helena'. She may have been unable to find that island because of an error in the assumed compass variation, leading to a mistake in the estimated longitude, and so sailed on without calling at any refreshment point on the voyage home.

MEANWHILE in the Netherlands, on 20 March 1602, while the Harmensz fleet was in the East Indies, the Dutch Government granted a monopoly charter to a number of merchant companies amalgamated as the Vereenigde Oost-Indische Compagnie (United East India Company — known by its initials, VOC) which would become internationally famous, and sometimes feared, for two centuries.

The charter conferred on the VOC exclusive rights to trade with the East Indies and other parts of Asia via the Cape of Good Hope or through the Straits of Magellan for twenty-one years. The Company could make treaties, raise armies, build fortresses, appoint officials and exercise almost unrestricted power in the name of the Republic. The VOC began with a capital of nearly 6.5 million guilders, a considerable sum compared with the £40,000 capital of the English East India Company, formed in 1600. The VOC's activities were destined to be profitable and successful enough to quickly elevate the young Netherlands nation to the status of a first-rate European power. It was a joint stock company with limited liabilities, and its former separate companies continued under various boards of directors, Amsterdam, Zeeland, Delft, Rotterdam, Hoorn and Enkhuysen, each of which sent out ships of its own.[17] Nominees to the board of seventeen directors (Heeren XVII) had to own shares to the value of at least 6000 guilders.

On 20 April 1602, only one month after the establishment of the VOC, a resolution of the Amsterdam Chamber declared an interest in 'some small yachts such as *Wachter, Duyfken* etc.', then away in the Indies, and suggested that if the admiral, Wybrant van Warwijck, and his Broad Council found any of these could be of useful service, they could purchase them at a mutually agreed fair price, which would be paid out of the returns of those fourteen ships on their return.[18] The VOC then instructed Admiral van Warwijck on the ship *Mauritius*, which would leave Texel on 17 June 1602 for Bantam, arriving there 29 April 1603, to buy these yachts, but only at a reasonable price.

Enlargement showing more detail of the ship *Duyfken* at Mauritius.

Duyfken was described[19] as of thirty lasten — a small cargo capacity — and sixty tons. She had already proven very effective in her role as an East Indies fleet messenger vessel capable of exploration in shallow waters while being a formidable little warship. A later record,[20] of 8 November 1603, showed that the deal was concluded:

> The yacht called Duyfken owes 2200 florins paid at the Receivers to Peter Evertsz Hulff, about the purchase of the aforementioned yacht and appurtenances for the purposes of this voyage [of the Steven van der Haghen fleet which sailed from Texel in December 1603] bought from him and received in accordance with the ordinate and receipt f2200-00-00.

In 1603, 2200 guilders for a *Duyfken* of thirty lasten was a fair price for an almost new ship.

On 18 December 1603 *Duyfken* set out again for the East Indies, this time in the expedition commanded by Admiral Steven van der Haghen, the first fleet wholly owned and equipped by the VOC. Willem Jansz was skipper of *Duyfken*.

From the Amsterdam Chamber came:
Geunieerde Provincien (United Provinces), of 350 Lasten, 700 tons, with Admiral Steven van der Haghen aboard.
Amsterdam (250 Lasten, 700 tons).
Gelderland (250 Lasten, 500 tons).
Hof Van Holland (garden of Holland) (180 Lasten, 360 tons).
Delft (150 Lasten, 300 tons).
Duyfken (30 Lasten, 60 tons).

From the Chamber of Zeeland:
Dordrecht (350 Lasten, 900 tons), with Vice-Admiral Kornelis Bastiaanz [also called Cornelis Sebastiaansz].
Zeelandia (250 Lasten, 500 tons).

From the Chambers of Hoorn and Enkhuysen:
Hoorn (350 Lasten, 700 tons).
Medemblik (125 Lasten, 250 tons).
West Friesland (350 Lasten, 700 tons).
Enkhuysen (150 Lasten, 300 tons).

Van der Haghen's twelve well-armed ships carried a total of 1200 'eaters' (crew members, merchants, soldiery and others) when they set sail. This was effectively a naval task force ordered to sail out and make war with any Portuguese or Spanish fleets encountered in the East Indies or on the way there and back. Some of these sailors were alarmed when they learned about this after Admiral van der Haghen opened his sealed orders at sea.[21] The survival rate of only one in three from previous voyages to the East Indies due to scurvy, malaria and various hazards of sea life, was bad enough without the added perils of armed conflict, and most had signed on as seamen, not soldiers. But it was too late now, despite their protests.

The admiral was instructed to take his fleet first to the Mozambique Channel, between the island of Madagascar and Mozambique, and lie in wait to plunder or destroy any Portuguese or Spanish carracks heading out from Lisbon. Next he was to move to Goa, the centre of Portuguese power on the western coast of the Indian mainland, to display the Netherlands flag and burn the enemy ships. Then van der Haghen was to go on to the east and sweep the Portuguese and Spanish out of the Spice Islands of Maluku.

After leaving Texel the fleet was separated at various times, and progress was slow. Apparently *Geunieerde Provincien* and *Dordrecht* were still sheltering at Plymouth in 1604 from 10 January to 12 February.

Then the ships pressed down in the Atlantic, south and then far south-west, close to South America, searching for the strong southerly winds. From 3 March to 30 May *Geunieerde Provincien* was reported to be at Majo, among the Cape Verde Islands off the northwest African coast, where *Dordrecht* and *Zeelandia* joined the fleet.

They rounded the Cape of Good Hope on 30 June 1604 and were at Mozambique on the East African coast probably from 28 July until 15 August.[22] A Portuguese carrack was captured and burnt, and a smaller vessel captured here was taken away and renamed as the *Mosambique*. Some more Portuguese carracks were expected soon, so *Delft*, *Duyfken* and *Enkhuysen* remained another five days after the other Dutch ships left, but evidently the Portuguese did not arrive, so they rejoined the fleet at the Comoro Islands north of Madagascar, about 12°S and 50°E.

Then the fleet sailed up to harass the Portuguese base at Goa on the western coast of India, from 29 September to 14 October, but with little result. They sailed on eastward for the Indies. After various shipping movements there, *Hof Van Holland* left the fleet early to go to Mauritius to search, unsuccessfully, for the missing yacht *Wachter* and then return home. No more was heard of *Wachter*.

Admiral van der Haghen in the flagship *Geunieerde Provincien* reached the Dutch base of Bantam, West Java, on 31 December 1604. *Duyfken*, *Gelderland* and the *Medemblik* had arrived there before him while *Zeelandia* had gone on to Cambodia. *Enkhuysen* would not show up until 21 August 1605, and she would be reported later as having run aground off Halmahera, in North Maluku, in 1607.

At Bantam the admiral was told that Fort Van Verre, which he had established on the island of Amboyna during his visit of 1599–1601, had been seized by the Spaniards. He was ready and confident to deal with this, but before sailing for Amboyna he appointed Jan Willem Verschoor as the VOC's chief merchant at Bantam.

After arriving in Amboyna van der Haghen wasted no time in opening an attack on the Portuguese. When asked what they were about, the Dutch informed the commander of the post that they had come to take possession of the island of Amboyna in the name of Prince Maurits, and henceforth demanded the surrender of the Portuguese fort. Within a couple of days, on 25 February 1605, the enemy surrendered, allowing the Dutch to take possession of the island and fort. On his way to the island, van der Haghen had captured a Portuguese ship laden with gunpowder and ammunition for the Portuguese fort on Amboyna. The quick surrender most likely had to do with the lack of ammunition.

The admiral then appointed Frederik de Houtman as Governor of Amboyna, the first territory the Dutch had established for themselves in the East Indies. His elder brother Cornelis de Houtman, aboard *Mauritius*, had commanded the difficult first Dutch voyage to the East Indies in 1595. In 1598 Frederik de Houtman as skipper of the ship *Leeuwin* had sailed with his brother from the Netherlands for Banda Aceh, Sumatra. Arriving there on 30 June 1599 they encountered some hostile Portuguese, and Cornelis was killed and his brother Frederik was taken prisoner for two years. Frederik de Houtman was released in August 1601 when another Dutch fleet arrived at Banda Aceh.

Frederik de Houtman was to serve as Governor of Amboyna for six years. In 1619, as commander of the ships *Dordrecht* and *Amsterdam* he would sail up the coast of Eendrachtsland (Western Australia). The Houtman Abrolhos islands acknowledge his visit. When the first Dutch ship, *Eendracht*, came to this coast in October 1616 her skipper Dirk Hartog and some of his crew went ashore at what is now known as Dirk Hartog Island, leaving an inscribed pewter plate nailed to a wooden post, marking the first known landing of Europeans in Western Australia.[23]

After the recapture of Amboyna on 25 February 1605, Admiral van der Haghen divided his fleet. He proceeded to Banda with the flagship *Geunieerde Provincien* and six other ships, including *Duyfken*, while Vice-Admiral Cornelis Sebastiaansz, with five ships sailed to Tidore. On 7 May Sebastiaansz' fleet 'sailed under foretopsails fair by the Fort and let fly with all their ordnance', according to the Englishman Captain Henry Middleton, who was witness to all the events.[24] It was not until 9 May that the Dutch prevailed, and then only because by some accident the Portuguese fort on Tidore blew up, killing some seventy people — at the very moment when the Dutch and Ternatean attackers were retreating.

Captain Middleton had arrived at Ternate on 21 May, and on 1 June 1605 he visited the King of Ternate with a letter from Britain's King James urging approval for an English post there. But the local Sultan Said was not impressed, complaining that they had done little since Sir Francis Drake's visit in 1579, so the sultan had written to the Dutch Prince Maurits promising exclusive trade with his Dutch friends and liberators, whose guns he now eagerly awaited.

Sultan Said seemed to be learning to use the arts of diplomacy, but a massive threat was looming from the north where the Portuguese were united with the formidable Spanish forces based in the Philippines. The Dutch were masters of mobile strategy and could choose lateral moves to achieve their objectives, but the Ternateans were no match for the imperial powers.

In September 1605 the fully laden *Geunieerde Provincien*, accompanied by *Duyfken*, sailed back to Bantam. The VOC orders provided now for the flagship to return to the Netherlands and *Duyfken* and *Delft* to remain in the East Indies for at least three years to search for new sources of trade. Before van der Haghen departed in October he placed *Duyfken* at the disposal of Frederik de Houtman, and sailed home safely to the Netherlands with his fully laden flagship *Geunieerde Provincien.*

Not long after the departure of the main fleet Willem Jansz, skipper of *Duyfken*, was ordered by de Houtman to sail to Bantam to obtain some very necessary provisions and ship's stores. De Houtman provided Jansz with orders to President Jan Willem Verschoor asking him to be of help to the skipper.[25]

A Voyage of Discovery

IN OCTOBER 1605 Captain John Saris, the English East India Company agent at Bantam, probably conducted his business aboard a ship from which he could conveniently observe activities of the rival Dutch East India Company. Saris had arrived at Bantam earlier that month and would remain there until October 1609, perhaps a good time to be away from the uncertainty at home. In 1603 Queen Elizabeth had died aged seventy after ruling England for forty-five years, a term which included the defeat of the Spanish Armada in 1588 when Britain and the Netherlands were close allies. She was succeeded by James I whose reign was to be marked by conflict with Britain's parliament and a preference for rapprochement with Spain.[26]

In November, 1605, Saris made the following entry in his journal:

> The eighteenth heere departed a small pinas [pinnace] of the Flemmings, for the discovery of the iland called Nova ginnea, which, as it is said, affordeth great store of gold.

Although Captain Saris referred only to 'a small pinas of the Flemmings', Dutch shipping records and other information clearly identified *Duyfken*. Saris was witnessing *Duyfken*'s departure on a voyage that would take the vessel and its crew to the shores of a land that hitherto had existed only in legend. The events of the voyage would have been recorded in the ship's log, which was still extant later in the seventeenth century; but what eventually happened to it is not known.[27] It was probably sent back on another ship to the VOC headquarters in the Netherlands for study by the Company's cartographers, as with Jansz' reports of earlier voyages.

Fortunately the discovery chart (or seventeenth-century copies of it) survived, and the 1606 voyage to Australia is confirmed beyond doubt by a series of later Dutch references. A good deal of information, written or inferred, can be gleaned from the chart, which is reproduced on page 41. To begin with, the inscription in the top right corner indicates that this map incorporates sections which had been previously surveyed:

This sea chart shows the way, both in sailing outward and back, followed by the yacht *Duijfien* [sic] when visiting the lands east of Banda unto Nova Guinea, but Guliguli on Cenam, and Ceram etc, is according to the drafting of Abraham Francken A. 1602. Done 20 April, and Nieuw Zelandt with the gounongapi ['fire mountain' — volcano] east of here has been sailed by Jasper Janssen de Jonge.

The remaining two-thirds of the *Duyfken* chart — from the Aru Islands eastward to the descending part of the New Guinea coast and then on to Australia's Cape York Peninsula, and the return to Banda — would have been ascribed to Willem Jansz in 1605–06. On her return voyage, *Duyfken* sailed up to a section of the (New Guinea) coast which was marked as 'Os Papuas.' The chart shows that the area intended to be explored was east of the Banda Islands

It is likely that the charts available to Captain Jansz included a 1587 map of islands of the western Pacific by Flemish cartographer Abraham Ortelius (1527–98). This map showed the Solomon Islands, discovered in 1568 by Alvaro de Mendana on a journey from Callao, Peru. But, in showing the nearby New Guinea and then Cenam and the 'Mal luc' Spice Islands, the Ortelius map placed the Solomon Islands about forty degrees, or 4400 kilometres too far east, and Ceram thirty-six degrees too far east. He drew Nova Guinea as a complete island, including a southern coast descending more than twenty degrees nearly 2000 kilometres south of the equator — more of a fat duck than the usual bird analogy. This duck belly sagged 1000 kilometres south of New Guinea's actual southernmost point of about 9°20′. The southern New Guinea coast on the 1587 map seems to be the convenient product of somebody's imagination, along with an ornamental scroll and snakes with goblin heads covering the top of a vague sprawl of land, presumably suggesting the (also imagined) South Land.[28]

Duyfken departed from Bantam on 28 November 1605, by the Gregorian calendar, but the chart of the discovery voyage begins from Banda, about 2700 kilometres further east. That far was known well enough by the Dutch, and the time was just about right for the seasonal north-west monsoon winds to carry the ship on its eastward course along the northern coast of Java, Madura, past the delightful island of Bali, and Lombok, Sumbawa, Flores, soon swinging north-east to return, in time for Christmas, to the familiar beautiful green Banda Islands.

Probably a week later *Duyfken* set out from the main island of Banda Neira and took a course slightly south of east. When they were about eight kilometres out, the supercargo Jan Lodewycksz van Roossengijn could look astern at the little volcanic Island of Roossengijn, whose name he had adopted when he came to the East Indies about six years before on another ship. He had wanted to avoid being confused with another Jan Lodewycksz already working in the Indies. That island is now called Hatta, but the mariners' hydrographic guide *Indonesia Pilot*, Vol. III, 1980, warns today of the danger of its off-lying 'Rosengain' Reef.

Duyfken sailed below some small islands south-east of Cenam which she had charted in 1602, and north of three more, the central one of which her chart called Thori (probably Tioor on today's marine chart). Then she turned further south for the Kei Islands,[29] to pass north of the main one, Kei Besar.

About ninety kilometres north of the Kei Islands lay what her chart called 'Nieu Zelant', probably a part of the Bomberai Peninsula of Irian Jaya. South of Nieu Zelant was marked as 'Marasich landt', appropriately referring to the marshy Adi Island there. In the distant north-east could be seen smoke drifting from the 655-metre 'fire mountain', the volcanic Gounongapi. Kei was referred to as Hooghlandt (high land), and Aru, about 130 kilometres east, as Larghlandt (low land). *Duyfken* seems to have called at Dobo on the north-western side of Aru Island. She cleared the top of Aru and headed east and south-east. Now began the specific task of locating and exploring the 'Nova Guinea' coast.

When Jansz inevitably found the coast about 440 kilometres east-south-east of Aru he did not name it but wrote his description, 'Larghlandt', and marked the landfall near a river inlet at 6°15′S on his chart. The descending shape of the coast here leaves no doubt about where the ship was heading.

Before them was the northernmost of three mouths of the main Digul River, a fearsome area to venture in any craft, let alone a square-rigged yacht, even several kilometres out to sea. The ship's boat would have been launched to survey the inshore area. 'Caution: Extreme caution should be exercised in approaching the coast near Sungai [River] Digul', warns *Indonesia Pilot*:

> The coast near the delta is low, muddy and covered with jungle which floods at high water and so there are few landmarks. The only distinctive

mark is a sandy beach about a quarter of a mile long near Tanjung [Point] De Jong where it trends SE. Neither this point nor Tanjung Modder ['Muddy Point', to the north] can be distinguished at a distance seaward'. [Such a strip of sandy beach would seem extremely rare on this muddy coast.]

The Jansz chart correctly described the De Jong Point location as 'Larghlandt'. South of it the land curved sharply back south-westward, and this part was called 'modder landt' (muddy land).

Indonesia Pilot warns of the Digul River:

Tidal Streams. Both ingoing and outgoing tidal streams may be strong. At or near spring tides, with the first rise after low water, a bore or flood wave about 2m to 3m high travels up the Sungai [River] Odammun. The wave in the form of breakers that could endanger small vessels.

Duyfken's chart suggests some navigational confusion around De Jong Point. One could only guess at the course taken, perhaps around past the northern entrance to Digul River and down to near the village of Tiuri and then doubling back well out, avoiding more hazardous shallows off what later Dutch explorers would name as Frederik Hendrik Island (now called Dolak). It would have been a difficult south-south-west course.

The name of 'Tiuri' was marked on the chart here, and then more low land. The ship sailed well out from shore, trying to avoid the shallows when she passed then un-named False Cape, shown today as Ujung Salah or Tg. Vals. In 1622 a Dutch map by Hessel Gerritsz would show an area near False Cape as 'Duyfkenslandt', and in the following year Jan Carstenszoon of the ship *Pera* would bestow the name of Caep Valsch (False Cape).

Frederik Hendrik Island is narrowly separated from the south-western coast of Irian Jaya by Selat Marianne (Princess Marianne Strait) and the low island, covered with dense forest, is so marshy as to be almost inaccessible. Some decades ago a wealthy American anthropologist named Rockefeller was reported to have been captured and killed by savage tribesmen in this area.

Indonesia Pilot cautions that False Cape, the south-west point of Frederik Hendrik Island, 'is dangerous to approach from west for at low water it is possible to run aground before sighting land'. The coastal bank on the south side of the island, it says, is very steep-to. It adds: 'Bad weather is usually experienced off False Cape. With any sea running, as is often during the NW monsoon, it is advisable to keep well away from the coast or a vessel will roll heavily'.

Duyfken crept past on a south-west course. The sounding lead showed shallow water continuing for more than 110 kilometres

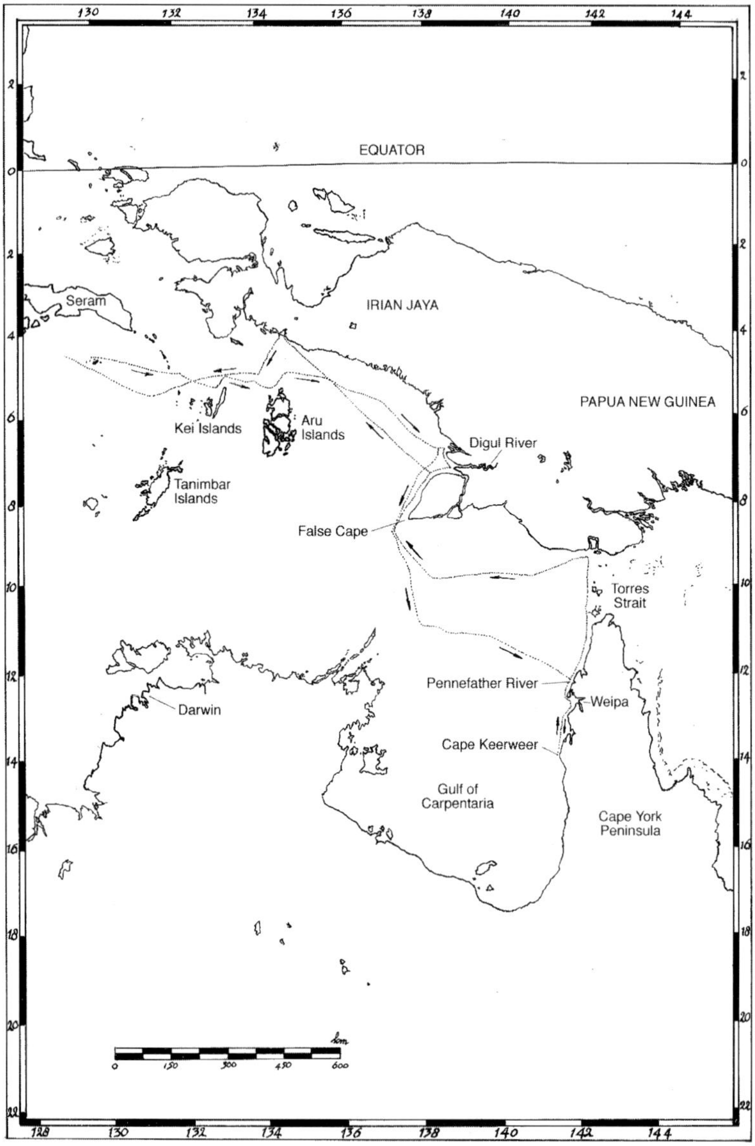

Reconstruction of *Duyfken*'s probable course on its 1606 discovery of Australia.

below False Cape to block eastward progress along the New Guinea coast Jansz had found. Then she swung to port and sailed about 190 kilometres south-south-east into the Arafura Sea before the favourable north-west to westerly winds. By now it was probably early in February 1606. With the strong monsoon driving her, the ship could speed confidently through the blue swells during the days and move cautiously under small sail at night. Much of the time it was a bit too deep to put out an anchor anyway. There were some shoals Jansz could have encountered if he had been on a more northerly course, and he could not know of the vast distant shallows in the east, around the northern extension of Australia's Great Barrier Reef.

Australia Pilot, Vol. V, 1972, published by the Hydrographer of the Royal Australian Navy, refers to the vast shallow bank known as the Arafura Shelf, belonging to the Continental Shelf surrounding Australia. A vigilant lookout should be kept when navigating in these waters. During December north-west to westerly winds spread, from west and north, across the Arafura and Timor seas and the Gulf of Carpentaria.

By January the north-west or west monsoon is established over most of the sea area north of 20°S and east of 110°E and has spread to the Torres Strait region. It continues to occupy all this area through February, but in March the winds become variable. By April the south-east wind has re-established itself.

The north-west or west monsoon is associated, especially at its onset, with considerable cloud and rain, and frequently with thunderstorms. These storms become more frequent, with heavy rain and squalls. The monsoon is most fully developed in January and February, when it extends farthest south and affects all parts of the Gulf of Carpentaria and the coast of Western Australia as far south as Exmouth Gulf (about 21°48′S).

In the transitional period following the north-west monsoon, generally of several weeks in March or April, the wind is unsteady but more often from north-west or south-east, generally light but not often calm except on the coast and then mostly at night. It seems that Willem Jansz had chosen the best time of the year for his voyage of discovery, both out and returning.

Approaching the Gulf of Carpentaria, his course was roughly east-south-east for about 320 kilometres before the ship swung to the south-east. After about 210 kilometres on this course there would have come a shout from the mast lookout: 'Landt in zicht'.

Taking shape was a long white beach with low sandhills and behind that, a featureless brown land of stunted greyish trees stretching back into the purple distance, very different from the beautiful lush colours of the volcanic islands of the East Indies. For the eyes of this crew it must have been a monotonous and uninviting view. But any land would be welcome now, seven hundred wet and windy kilometres and perhaps a week after their last glimpses of 'Nova Guinea' at its difficult southern corner of False Cape.

Now a river mouth was to be seen beyond the port bow, and from the masthead the lookout could spy a bright yellow sandy lagoon with the river soon widening and stretching back several miles inland. There was a promising clump of trees on the north side of the entrance: a chance of getting some dry firewood for the galley stove. The cook needed some wood that would burn in this hot but damp climate.

The leadline revealed gradual shallowing. About four nautical miles out, a thirteen-metre bottom was close enough in these circumstances for *Duyfken* with her draught of 2.1 metres. Sails were furled, the anchor splashed down and disappeared in the pale brown water, and the ship's boat was made ready for launching.

An officer and half-a-dozen sailors would have comprised the landing party. Their routine task was to search for water and firewood, make an initial survey of the river and try to question any local inhabitants about trade prospects for the Company. But they should return before sundown because the ship would have to move out to the safety of deeper water for the night. In the quietness of the skipper's cramped cabin Willem Jansz studied the chart. Later, he agreed when the uppersteersman dipped a quill pen in the inkwell and wrote confidently: Nova Guinea.

But *Duyfken* had discovered Australia. The shape of this chart was the enduring evidence.

The ship's boat, loaded with the usual water cask and a few provisions, and another cask to be filled if possible, was lowered and cast off, drifting quietly shoreward as the sailors fended off the ship and began rowing. The boat mast went up, the oars were shipped and soon the sail bloomed. Aboard ship the watchers tried to keep the boat in sight as it glided into the distance under the glaring sun until only the pale brown sail could be seen fluttering gently against the land, and then it too disappeared.

The boat crew carried some arms, having been customarily warned to be vigilant at all times for the danger of ferocious tropical creatures, and also the dreaded head-hunting Nova Guinea savages. With muskets, swords and pikes, they were prepared for trouble but under the sort of tropical downpours that had overtaken the ship since they had last sighted land, it would be difficult to keep gunpowder dry. It was about the middle of the north-west monsoon season.

At midday the navigator undertook the ship's regular procedure for calculating latitude by shooting the sun — measuring the angle between the horizon and the sun. He would have used the cross-staff, or possibly the more advanced back-staff invented about 1594 by the English mariner John Davis. This latter had the advantage that the observer did not have to look directly into the sun. There was an astrolabe for measuring the altitude of stars, and it could also be used for the sun.

The first Dutch sight of Australia: *Duyfken*'s 1606 mast-height view of Pennefather River, Cape York Peninsula.

The ship's position, off the mouth of what clearly represented today's Pennefather River, was marked at about 11°48′S on *Duyfken*'s chart. This was an error of forty-eight kilometres: the actual latitude here is 12°14′. No longitude (east–west position) was shown; a successful chronometer that would maintain the necessary highly accurate time at sea to enable calculation of longitude would not be available until 1761.

In the late afternoon, after the ship's boat returned from shore, *Duyfken* headed back out to sea for the night. The officer from the boat reported to Captain Jansz, and the uppersteersman wrote on his chart: 'R. met het Bosch' — River with the Bush.[30] Now there was some suitable firewood for the galley stove, brought back by the first whites recorded as having visited the unrecognised continent.

Whether the sailors saw any of the local inhabitants this time on the apparently silent land is not known. But it could well be

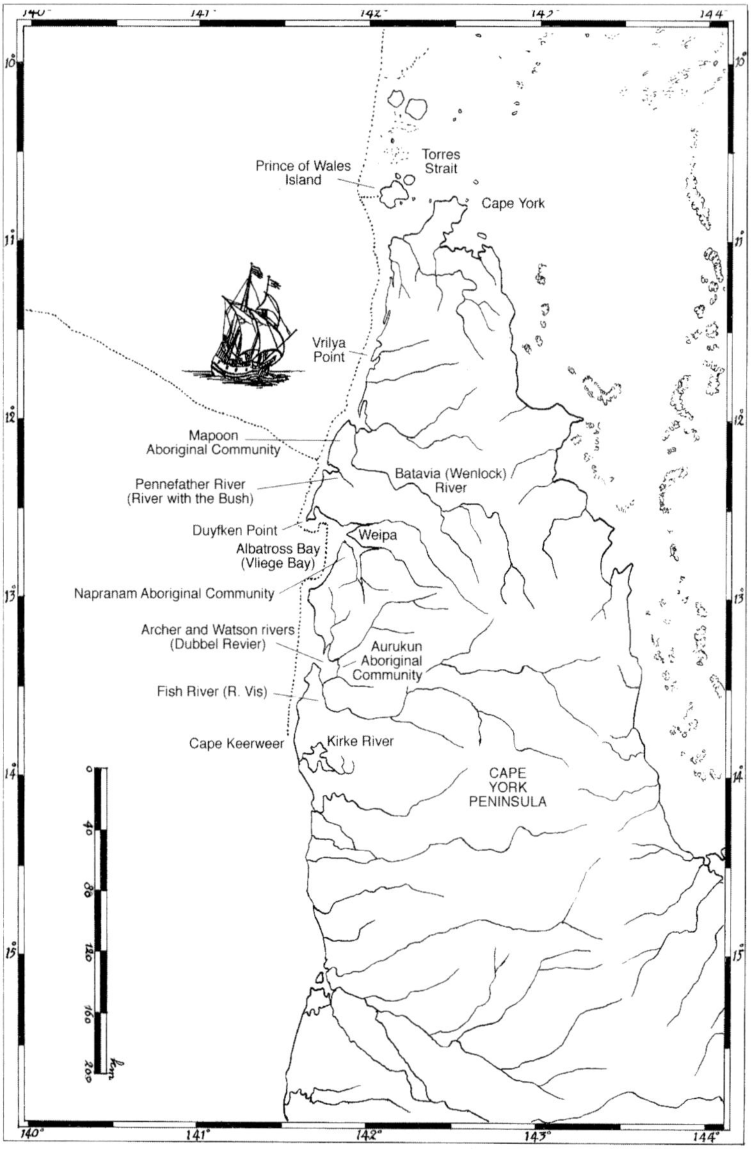

Detailed representation of Jansz' course along the coast of Cape York Peninsula.

assumed that the visitors were quietly watched by the people whose ancestors had occupied the land for at least 40,000 years, perhaps beyond 100,000 years. They traditionally built their own canoes, some with outriggers,[31] although they were generally smaller (between 4.5 and nine metres long) and less well finished than those of the Torres Strait Islanders. But the ship *Duyfken*, a little more than twenty metres long, was probably in the eyes of these Aborigines a great brown log with several bare limbs reaching up to the sky — but this had to be too big for a tree. Surely no tree could grow to this size, the warriors would exclaim. The elders, custodians of countless centuries of Dreamtime wisdom and law, a comprehensive religion covering the past, life and death, could not have explained it either.

Heading south about six kilometres from shore, thirty-eight kilometres south of 'River with the Bush', *Duyfken* rounded a point which her chart showed at 12°8′, but again was not named. The modern chart shows Duyfken Point about 12°34′ — an error of about forty-six kilometres. This feature — actually four low sandy points covered with thin casuarina trees — was named by the British explorer Captain Matthew Flinders on board HMS *Investigator* in 1802, as a tribute to Willem Jansz and *Duyfken*. Jansz would have given its inshore and offshore shoals a wide berth before then venturing cautiously into a wide inlet which he named Vliege (Fly) Bay, probably in response to irritation from the bush flies or, more likely, swarms of midges from the nearby marshes.

The ship's boat surveyed about forty kilometres into where the inlet is joined on the northern end by Mission River which runs past today's Weipa bauxite mining industry centre, and at the southern end, the Embley River. With their ship at least thirteen kilometres away, safely outside the wide shallow bay's ten-metre bank, the crew probably spent at least one night either uneasily ashore or in the boat at anchor, wondering about the floating logs and unpleasant bumping and splashing noises. The excellent shape on the chart, resulting from their exploration, is unmistakably that of Albatross Bay.

Clearing Boyd Point at the bottom of this bay *Duyfken* maintained a safe course about ten kilometres from shore while rounding Pera Head and Thud Point and passing False Pera Head. Now the land sloped away increasingly to the east, but the chart showed a straight course south, and the boat crew could have remained aboard ship for the survey of about twenty kilometres until another inlet could be spotted from the masthead.

Here the ship's boat had to be sent in through the shallows surrounding Worbody Point — only half-a-metre at low water. Even this craft risked being stuck on the muddy bottom approaching the entrance to the junction of several rivers — the Ward, Watson, West Archer and Archer.

About four miles up the watercourse, leading to the long, winding and branching Watson River, was a sandy beach. Around here today is the big Aboriginal settlement of Aurukun. The Dutch sailors may have seen some Aborigines watching apprehensively from the banks of these meandering rivers, probably warned by relayed smoke signals that strangers were approaching.

The boat crew might well have been apprehensive when surveying this isolated area. Their ship would have remained outside the eight-metre bank about fourteen kilometres out to sea from the river mouth — well out of sight from the river position, and this survey could well have led the boat party to make more landings to look for fresh water and for trade possibilities, the object of their explorations. It was again a good survey to achieve the results as drawn on the *Duyfken* chart.

The river was shown with the entrance about 13°8′S — a navigational error of only twenty-four kilometres. It lies on the modern chart at 13°21′S. Willem Jansz appropriately named this feature 'Dubbel Rev' (Double River): parts of the Ward, Watson, Archer and West Archer rivers would be seen from the boat.

Leaving Dubbel Rev at Worbody Point, the ship's boat had to creep about eight kilometres out to sea to clear the mangrove marshes and the worst of the shallows, and then continue southward following the ship, which could be seen about another eight kilometres out in the deeper water. She would still have been kept prudently outside the eight- to ten-metre bank. In the continuing shallows inshore, the seawater had taken on an eerie yellow colour. Along the shore as far as the eye could follow there stood an almost continuous rocky wall with pretty strata of pink, white and yellow: bauxite deposits, but that would mean nothing to these explorers. Behind that wall, into the distance rose the same mass of stunted grey-brown trees.

About sixteen kilometres south of Dubbel Rev was another inlet. The ship would still have been at least eleven kilometres out now, so the boat sailed in through the bright green mangrove banks and followed the river as it led south.

A few miles upriver was an opportunity for the boat crew to cast fishing lines and catch as many fish as the boat could carry. When they returned to the open sea their ship would make a welcome sight about eight kilometres out under the late afternoon sun. After the tired boat crew were back on board with their cargo and survey report the uppersteersman wrote on his chart 'R. Vis' (River Fish) at about 13°18′S.

What must be *Duyfken*'s River Fish is shown running from south to north at 13°30′ on the June 1992 Royal Australian Navy Hydrographic Service chart Aus 302, but it is not named. The Jansz chart showed it about 13°18′, only nineteen kilometres north of this latitude. The modern chart warns mariners that shallow coastal waters south of here, down past Cape Keerweer, are inadequately

Duyfken Point, named by Captain Matthew Flinders, RN, 1802.

surveyed or unsurveyed. Most likely the name River Fish celebrated a good haul of barramundi, regarded widely as Australia's most delicious fish, which had been plentiful in the rivers of Cape York Peninsula in the wet season.

Duyfken continued a southern course for about thirty-five kilometres to where the coast was seen to fall away to the east. Here, at about 13°40′ on his chart, Willem Jansz or the uppersteersman wrote 'Cabo Keerweer' (Cape Turn-about), and drew below it the coastline still running southward about as far as could be seen from the masthead. The modern chart shows the almost imperceptible Cape Keerweer another twenty-five kilometres south of the position shown on the *Duyfken* chart.

One might expect that it was an event of some significance which led to the decision to turn back and thus the naming of

Cape Keerweer. Possibly it was the loss of several of the crew of the ship's boat. In a later chapter the Aboriginal elders' handed-down, unwritten version of their warrior ancestors' fatal encounter with *Duyfken*'s boat crew nearly 400 years ago may answer some of the questions.

Duyfken turned back northward, probably followed by the same sequence of smoke signals that had pursued the ship down the coast from Pennefather River. She retraced her voyage to 'R. met het Bosch' (now the Pennefather River), nearly 210 kilometres north by Captain Jansz' reckoning but actually little more than 180 kilometres. Perhaps because he was now battling northward with contrary and changing winds he felt he had travelled the greater distance.

Then, instead of returning into the north-west where he had come from, Jansz continued his northward survey of the 'Nova Guinea' coast. At a little more than thirty kilometres, a clump of high trees probably indicated another inlet, today's Cullen Point and the entrance to the Wenlock (formerly Batavia) and Ducie rivers, drawn on the *Duyfken* chart about 11°27′S. The correct position of the entrance is about 11°57′. Cullen Point would have been concealed by the thick mangrove fringe in the shallow water. The ship would have anchored safely about ten kilometres out where this estuary, almost surrounded by mangroves, could be observed from the mast top. The landing boat, dispatched to survey the inlet, had to sail carefully in from the north-west, crossing a bank of about 0.3 metre and picking a way through depths of less than a metre to get into the southern channel of the Batavia River, as it would be named in 1623.

The *Duyfken* chart showed just one broad river, about eight kilometres wide, running into the sea from about thirty-five kilometres east-south-east. Beside its open eastern source was written 'Larghlandt'. Somewhere on the Batavia River the *Duyfken* boat crew had a fatal encounter with Aborigines that would be referred to by the next Dutch visitors to this coast in 1623 and by others later.

The boat could have followed the doubtful Batavia River course south through the surrounding thick dark green mangroves for several miles before reaching a sandy beach on one bank, marshes on the other.

Probably on this sandy beach, south of Cullen Point and on the western side of the Batavia River, today's Mapoon, *Duyfken*'s boat came ashore on the beach — the only likely landing spot — and ran into trouble with the local inhabitants. This encounter with Aborigines was referred to in documents quoted by J. E. Heeres in *The Part Borne by the Dutch in the Discovery of Australia 1606–1765,*

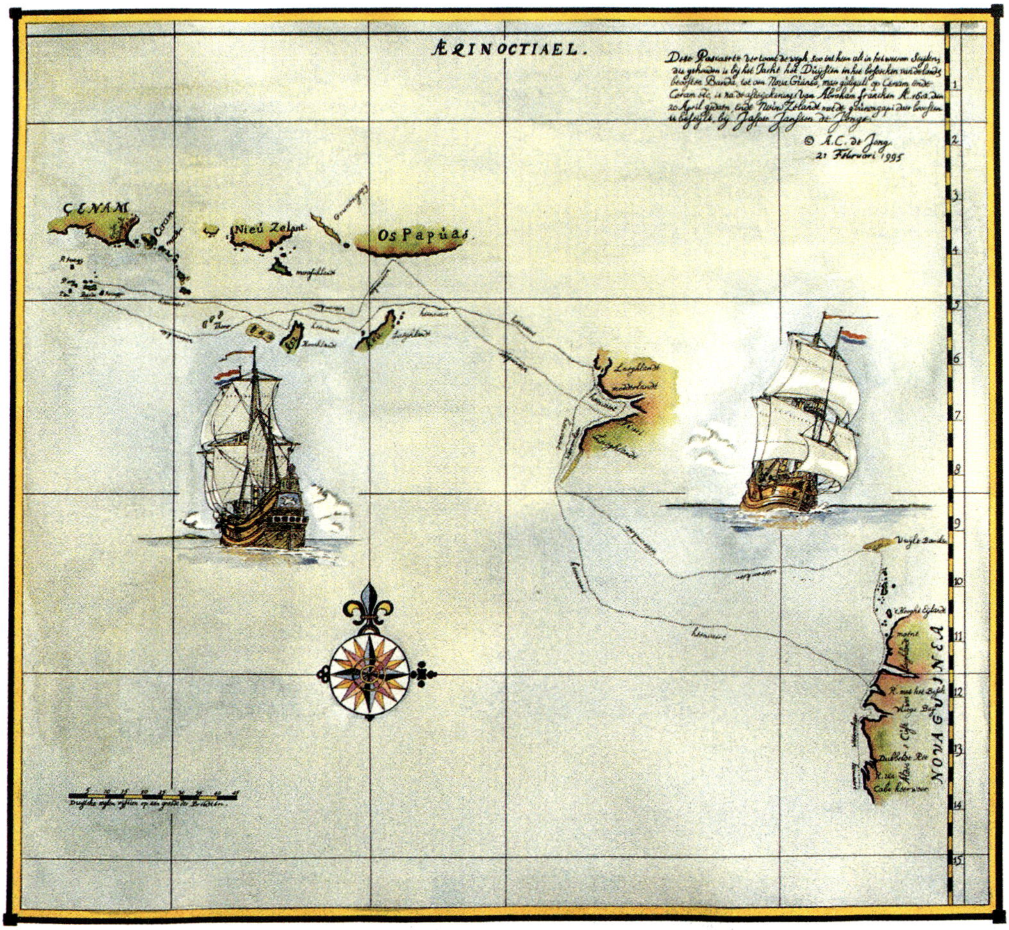

The Willem Jansz chart of 1605–06 showing *Duyfken*'s course on its retrospectively momentous voyage to Australia. Compare the identification of landfalls with the map on page 32. (Copy of seventeenth-century chart held by the Mitchell Library, Sydney, with colour and ship images added by Adriaan de Jong.)

published by the Royal Dutch Geographical Society in 1899[32] with details from the log of the VOC ship *Pera* which departed with *Arnhem* from Amboyna in January 1623 for 'Nova Guinea'.

The Ducie River does run from a marshy south-east source about thirty-five kilometres inland. But it is unlikely the boat could have found enough water under her keel to proceed that way after about eight kilometres of laborious navigation among the mangrove islands which blocked much of this passage of frequently less than one metre depth. Perhaps the shape of the river has changed since 1606 or, as it appeared to me, its upper reaches were the result of distant guesswork by a demoralised boat crew retreating from the Batavia River.

Duyfken continued the voyage northward from Cullen Point, now presumably heading several kilometres farther out to sea, to miss a sand spit and shallows reaching out about eight kilometres toward Kerr Reef (depth 1.2 metres). A clump of trees about twenty kilometres up the coast indicated another river mouth which would be that of the Skardon, but the river was not entered.

North of here the outline of the *Duyfken* chart is difficult to follow for a while — possibly because the ship was further out from shore and the remaining boat crew were unhappy about going in and risking more hostile receptions, but the chart suggests that at least one more river mouth, probably the Jackson, was spotted.

At about 11°S *Duyfken*'s chart shows a projection which would be Vrilya Point — actually at 11°14′; latitude was getting more accurate again! Probably the rough and wet weather in February and early March had affected some measurements of latitude when it had been difficult to make a clear sun shot at noon.

Soon after this Jansz wisely (or by good luck) swung the ship out to a north-north-west course to pass west of what would have been Crab Island, off Slade Point, about twenty-five kilometres north of Vrilya Point. On the chart's coast nearby was written the word 'Moent', but it is not clear what the word means and whether it applied to the claw-shaped island. *Australia Pilot* warns that extensive sandy shoals extend more than twenty-seven kilometres west of Slade Point (10°59′S).

As *Duyfken* was leaving the coast that her chart named as Nova Guinea, Jansz or the navigator wrote (from south to north) a general description on the chart: 'All this coast was low land'. By now the ship was too far away for them to observe the line of hills running up from Bamaga to the top of the peninsula.

Duyfken had lost at least one of her crew in the encounter on the Batavia River, perhaps another three or more at Cape Keerweer. Willem Jansz might well have been convinced that no good was to be done for the Company at this wretched place.

North of Vrilya Point they probably sailed further seaward than their chart indicated, clinging to the safety of a ten-metre depth. It was a hazardous stretch across the western exit of Endeavour Strait/Torres Strait but the competent surveying went on. Within a wide area of great sandbanks, three little islands were drawn.

These would be Woody Wallis Island and Red Wallis Island; the third was perhaps merely a high part of Rothsay Banks, but taken to be an island.

By now it was nearing the end of March 1606, with the weather beginning its change into a period of unsteady wind direction before the south-east monsoon season. The mast-top lookout spotted something in the distance to starboard, a hilly island about twenty-four kilometres away. Water and firewood would have been running out, and Jansz apparently decided to examine the prospects in the east.

Evidently the full south-east monsoon had not yet arrived. A dotted line on the chart showed where the ship sailed a little south of east toward the middle of 'de Hooghe Eylandt' (the High Island — undoubtedly our large Prince of Wales Island with peaks of up to 246 metres) to his estimated position of 10°30′. This was very close to the correct latitude. Probably the ship's boat went ashore and some of the crew climbed the highest peak to study the shape of the several other islands nearby. Jansz reversed course westward and then continued north to plot more islands including Badu and Moa and half-a-dozen little ones. Apparently he missed Booby Island, not far to the west.

Continuing his almost northward course he found himself in a worsening nightmare of shoals. Nearly 110 kilometres from the shores of 'de Hooghe Eylandt' (and the northern tip of Cape York, which he had not seen) he could have anchored the ship many times and sent off the ship's boat in various directions to where the crew could step out of the boat and wade in the shallows. Eventually, about 9°15′ on his chart, he faced a reef strip west-south-west to east-north-east, more than seventy-five kilometres wide, an absolutely impassable bank.

The *Duyfken*'s chart showed this one of many such obstacles blocking northern progress to Nova Guinea as 'Vuyle Bancken', meaning 'dirty, foul sandbank or reef area'. What looks like part of the same 'Vuyle Bancken' is seen on the RAN Hydrographic Service chart Aus 376, Torres Strait. It refers to 'Discoloured Water' around latitude 9°40′S and longitude 141°44′E, with depths of 0.6, 1.2, 1.5 and two metres.

Jansz had to turn back to a west-south-west course. After about thirty-five kilometres, the sight of a narrow sand strip on his starboard bow (Turu Cay) would have persuaded him to maintain this west-south-westerly course, and after another fifty-five kilometres before the increasing south-east breezes of April he may have noticed Deelder Reef with 0.9 metre of water, possibly warned by white water breaking over it. Finally, he was confident enough to turn north-west for about 320 kilometres to reach the position where he had left the shallow bank south-west of New Guinea's False Cape.

After a couple of days on a north-north-westerly course *Duyfken* turned in toward the coast of Frederik Hendrik Island to anchor and send in the boat for urgent water and firewood.

Jansz, or the navigator, wrote on the chart 'Tiuri', for the village there. It seems likely that there was an encounter with the hostile indigenous people of this area, bringing *Duyfken*'s loss of crew to the reported total of nine.[33]

Banda lay another 1080 kilometres to the west-north-west, and *Duyfken*, despite her depleted complement, inexplicably took a more northerly course, to the section of New Guinea coast above the Aru Islands, drawn vaguely on the chart and called 'Os Papuas', where it may well have been impossible for the ship's boat to reach this muddy shore below the mountain range with its snowy peak of nearly 4000 metres. Perhaps the Os Papuas section had been charted previously.

Duyfken turned south-south-west for about 150 kilometres, but the low-lying Aru Islands could not be seen. She swung sharply back westward to pass probably within distant mast-top sighting, about twenty-four kilometres, of the 'Hooghlandt' Kei Islands but without calling there. Fortunately it was then familiar sailing below Thori (Tioor) and the other Watubela Islands and then west-north-west to the south of Banda Islands.

About the end of April 1606 *Duyfken* dropped anchor at the conclusion of her unrecognised but momentous discovery voyage. Willem Jansz and the chief merchant officer Jan Lodewycksz van Roossengijn soon went ashore with the ship's log, to make the most of a dismal report to the VOC governor about their costly trip to 'Nova Guinea'.

The Indian skipper of a Java junk, laden with mace and nutmegs, saw *Duyfken* lying at anchor there and made some inquiries before he set off for Bantam. There, on 25 June 1606, he would be questioned by Captain John Saris about various shipping activities he had observed around the Spice Islands. So Captain Saris penned in his

English journal the first ever written entry relating directly to the Dutch discovery of Australia, although described as Nova Guinea:

> The fifteenth [the 25th by the Gregorian calendar] of June here [Bantam] arrived Nockhoda [a skipper] Tingall, a Cling-man [of the east coast of India] from Banda in a Java Juncke laden with mace and nutmegs, which he sold to the Guzerats [traders of the western coast of India].
>
> He told me that the Flemmings Pinasse which went upon discovery for Nova Ginny was returned to Banda, having found the iland: but in sending their men on shore to intreate of Trade, there were nine of them killed by the Heathens, which are man-eaters. So they were constrained to returne, finding no good to be done there.

Duyfken was to sail into more adventures before her end, and her skipper Willem Jansz had a distinguished career ahead of him yet. But in the meantime, more Australian maritime history was being made in 1606, this time by Spanish captains whose contribution was also to remain unrecognised, ignored or delayed.

On the night of 3 October 1606, about eight months after Willem Jansz and *Duyfken* had arrived at the Pennefather River on Cape York Peninsula, Luis Baez de Torres' galleon *Almiranta* and Don Diego de Prado y Tovar's smaller vessel *Three Wise Kings* were anchored over what is now known as Peebles Shoal in Endeavour Strait between Cape York and Prince of Wales Island. They were within several kilometres — obviously in clear sight — of the tip of Cape York. In the falling tide the bottom of *Almiranta*'s hull began bumping on the shoal, and in the emergency Torres probably launched the ship's heavy boat and jettisoned a cannon, and paid out more anchor cable to move into deeper water without suffering any damage.[34]

Torres and Prado had sailed from Peru in a faltering, ill-led expedition of Pedro Fernandez de Queiros in search of the 'South Land'. Queiros decided he had found it when they reached the northern islands of the New Hebrides, and named it Tierra Austrialia del Espiritu Santo — South Land of the Holy Ghost. He sailed back to Peru but Torres and Prado, unconvinced, continued west-south-west and north to reach the eastern cape of New Guinea before turning south-west on the nightmarish search through shallows and reefs for a westward passage.

After traversing the strait that connects two seas — the Coral and the Arafura — they turned north for New Guinea and eventually the Philippines. Torres wrote to King Philip III about his exploration but his letters were filed in the archives and not disturbed until 1780.

IV Further Adventures

WHILE *Duyfken* was busy with her south-eastern discovery voyage a number of significant events were occurring in the Spice Islands. One morning about the middle of March 1606 there was an eclipse of the moon, which was interpreted as an omen of impending calamity for the sultan and people of Ternate.

Sure enough, on 26 March 1606 a huge Spanish and Portuguese invasion fleet of thirty-six vessels arrived at Ternate, carrying more than 3000 men commanded by the Philippines' Governor-General Don Pedro de Acuna. They soon overwhelmed the fortress, and the sultan, his family and some supporters fled to Jailolo on the big island of Halmahera.

On an assurance of safe conduct he returned to Ternate to sign a treaty of capitulation to the Spaniards, who claimed the island in the name of the King of Spain. They also garrisoned the Portuguese Fort Gammalama with Spanish troops, restored the church of Sao Paulo and returned it to the Jesuits for their missionary activities, converted the principal Ternate mosque to the Jesuits and named it the Convent of San Francisco.

In November 1606 the Ternatean leaders signed an agreement to recognise the King of Spain as their lord, not on pain of death to purchase any arms without Spanish approval, not to trade with anybody but the Spaniards and Portuguese and their friends; to return all gold, silver, jewels and religious items to the island of Ternate, and not to trade or speak with the Dutch, on pain of death and confiscation of goods. Sultan Said and his leaders were exiled to Manila. The rulers of Ternate, Tidore, Bacan and Siau all swore allegiance to the King of Spain and promised not to admit the Dutch to their lands but to bring all their cloves to the King of Spain or his representatives.[35]

It was a glorious victory for the Spaniards, to be seen as an indication of divine blessing upon their labours. But the Dutch were well established and prospering in south Maluku. They would be back before long, and then these Malukuans would see that they had an ugly side too.

Soon after *Duyfken* returned to Banda, her skipper Willem Jansz was given command of the much bigger ship, *West Friesland*, which had left Holland on 18 December 1603, with a total of 130 persons aboard, and arrived at Bantam on 31 December 1604. Some VOC papers relating to activities at this time are missing, but the circumstances suggest that Jansz was given command of *West Friesland* as a reward for his competent service on *Duyfken*, even though no good was likely to be done for the Company at Nova Guinea.

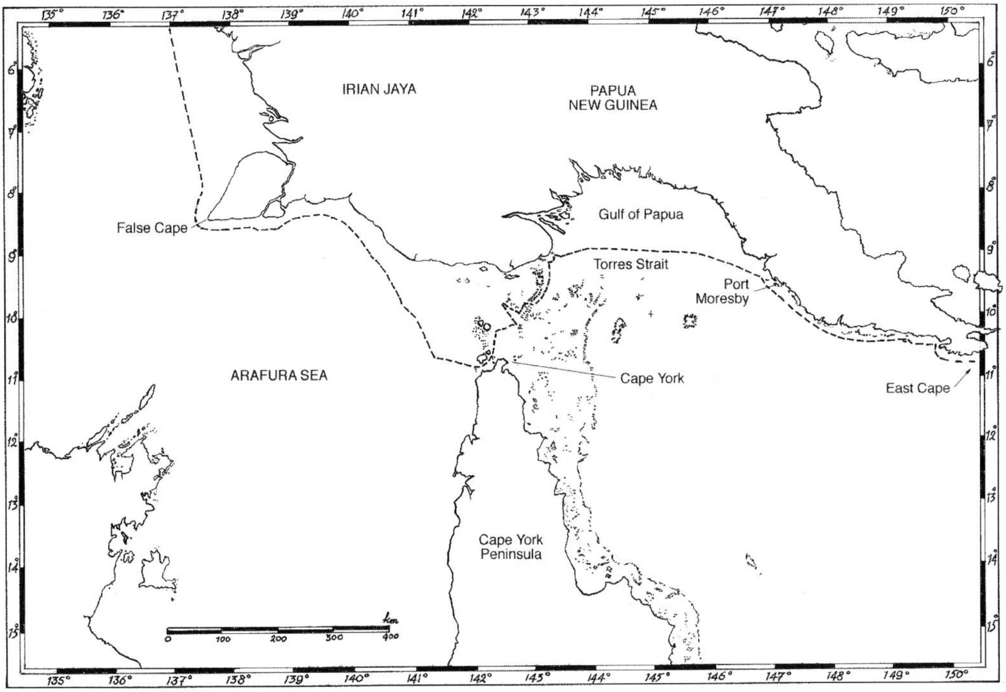

Torres Strait: part of the course followed by Torres according to Hilder.

On 4 March 1607 Paulus van Solt, supercargo on a Dutch ship (unnamed) which visited Amboyna, wrote the following report in his 'Narrative and Journal of the voyage made from Bantam to the coast of Coromandel and other parts of India':[36]

> ... through God's mercy [we] arrived before the Castle [Fort Victoria, Amboyna] ... here we found ... the yacht *Duyfken*, which had come from Nova Guinea.

Nothing in the records indicated whether van Solt's brief report referred to the *Duyfken*'s voyage of 1605–06 or a later one, perhaps to what actually was New Guinea. It seems unlikely that she would have made another voyage to Australia, but anyway, she was soon reported to be back in the action.

Later in March a fleet of five ships, *Oranje*, *Mauritius*, *Erasmus*, *Enkhuysen* and *Kleine Son*, under command of Cornelis Matelieff, arrived at Amboyna, where they found the yachts *Medemblik* and *Duyfken*. On 3 May Matelieff set sail for Ternate with the above vessels, planning to resume the offensive against the Portuguese, and on 13 May, when nearing Ternate, *Medemblik* left the fleet to sail on alone for Bantam.

But Ternate had been recently reinforced by Spanish forces from the Philippines, and when the Spanish captains Torres and Prado had departed on 1 May for Manila, they left behind their launch *Three Wise Kings* and about twenty Spanish soldiers to help defend the Portuguese fort Gammalama against their common enemy, the Dutch.[37]

When Admiral Matelieff became aware of the strength of the enemy fortifications on Ternate he decided to establish a fort and garrison there near the village of Malayo. He appointed Gerrit Gerritsz van der Buys as 'Captain of the Castle' and Jan van Roossengijn as 'Captain of the Naval Force', which consisted of the ships *Enkhuysen*, *Kleine Son*, *Delft* and the yacht *Duyfken*.

By 8 June 1607 the Ternate castle was formally established as Fort Oranje, where the local chiefs could take refuge from the Spaniards. A new Dutch treaty with the Ternateans was signed, securing the clove monopoly, while the Spaniards continued to threaten the post with constant reinforcements from Manila.

Matelieff then sailed to China, but this yielded little profit so he returned to Bantam in November 1607 to meet the fleet of Admiral Paulus van Caerden, whose next orders were to conquer Malacca, on the western side of the Malay peninsula. Matelieff was not in favour of an expedition just then against Malacca because of the western monsoon, and he advised van Caerden to go to Ternate to expel the Spaniards from there.[38]

Meanwhile the Portuguese had stormed Ternate's Fort Oranje on 15 July 1607, but the Dutch with local help managed to hold out and continued for the next several months to consolidate their risky position. Roossengijn sent *Duyfken* out for urgently needed supplies. At Bantam they wrote:

On 6 November [1607] the yacht *Duyfken* has left Bantam for Ternate, loaded with 10 last [weight-volume measurement] of rice and 7 leggers [casks] of Arrack, and is ordered to sail to Gresik [on East Java near Surabaya] in order to get from there a full load of rice and beans, and after this to sail as quickly as possible back to the Moluccas [Maluku].

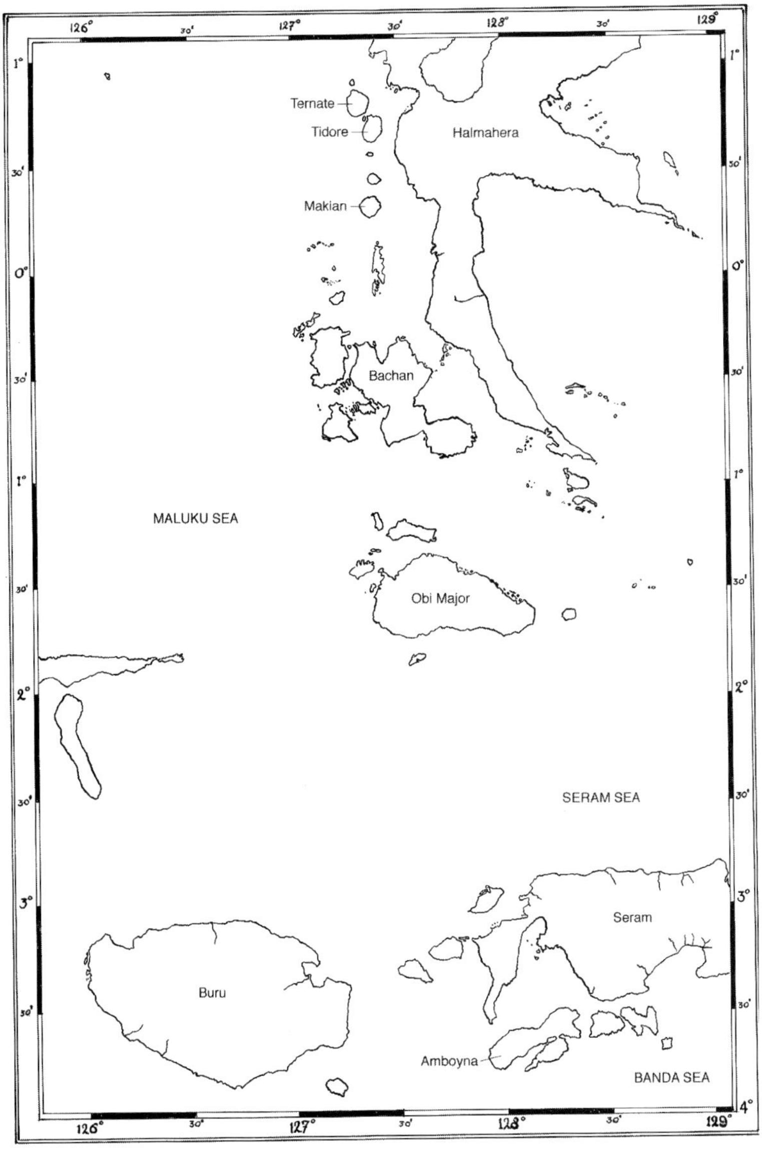

The Maluku islands and Amboyna.

On 24 November 1607 Admiral Matelieff arrived at Bantam on the ship *Oranje* with his fleet and found Willem Jansz, whose ship *West Friesland* had been disabled and abandoned near Mauritius early in 1607 on the way home to Holland. The crew had been saved but the cargo was lost. Jansz and the supercargo of *West*

Friesland, named Olpherts, were picked up there by a skipper, Jan Fransz, and another supercargo named Claesz Simonsz and taken back to Bantam on the yacht *Madagascar*.

In December 1607 *Enkhuysen* ran aground on a beach at Tuban on the northern coast of East Java and was lost. The cargo was salvaged by the ship *Delft* which was unable to sail back to Ternate, perhaps against the north-west monsoon by now. Roossengijn recorded his suspicion that this skipper was fearful of a deteriorating situation at Ternate, with the threat of more enemy reinforcements arriving from Manila, and was avoiding his responsibilities.

On 6 January 1608 *Madagascar* was back in Banda, but after a few days she returned to Amboyna where she was found to be in such bad condition that she was condemned as a wreck.

The Dutch were running short of food, but *Duyfken* fought her way through enemy naval forces to arrive at Ternate on 20 January 1608. Roossengijn, in a letter written aboard *Delft* in May 1608, described some events which followed:

> 29 January, I sailed off with the aforesaid yacht [*Duyfken*] and captured a prahu which came from the island of Makian and was on its way to Ternate with about 2.5 last of cloves, about which we came to blows with the enemy next day with two galleys and a fregatta [smaller type of galley] for about five hours, and they fired more than 80 shots at us, but God be praised no injury was done to us.

On 20 April 1608 Willem Jansz became skipper of the yacht *Delft*, which had stayed in the East Indies after arriving late in 1604 with *Duyfken*. The merchant Jan van Roossengijn, who had been with Jansz on *Duyfken*, was now aboard *Delft* with Jansz when they sailed with Admiral van Caerden to Ternate. On 21 June 1608 Jansz was actively involved on *Delft* with the capture of Makian Island.

Admiral van Caerden finally arrived at beleaguered Ternate on his flagship *Banda*, with several of his seven ships which had left Holland in April and May 1606 with a total of 1060 men on board. The other six ships of van Caerden's fleet were: *Bantam*, *Ceylon*, *China*, *Patania*, *Ter Veere*, *Walcheren* and *Zierikzee*. The last of these had run aground and was lost off Mozambique after the rampaging force began its operations against the Portuguese base there.

Van Caerden had orders to capture the Portuguese fort Gammalama on the western part of Ternate, and also the nearby island of Tidore. The Ternateans had promised to provide military assistance to the Dutch. Van Caerden reported later on how the events unfolded:

> ... We departed from Malayo [port on the south of Ternate], to Tidore, arriving there on the sixteenth Junio [June 1608] and anchored near the old castle ... having inspected twice where we might be able to land, then following the inspection and careful consideration by the Council [it] is decided not to land as it appears in the book of resolutions including the reasons to which I refer.
>
> After that, in accord with what was decided by Council, we sailed away from Tidore with the ship *Patania* and the yachts [*Delft* and *Duyfken*], leaving the other ships there, taking with us most of the corre-corre [these were very fast Maluku fighting boats, with double outriggers and central platform and powered by sails and oarsmen], and we sailed to Makian before the town of Taffaso where we arrived on the 20th [June 1608] ...
>
> ... on the 21st [of June 1608] in the morning [we] landed and captured [the town of Taffaso, on Makian Island] by force without much injury to our people though it is very much a stronghold and if there had been 30 or 40 Portuguese I am afraid we would not have captured this place.[39]

After van Caerden's force conquered the Spanish fortress on the island of Makian the spoils of this victory were reported to include some cannon, 'also 36 swivel cannon, two falcons and a camel,[40] 24 gongs and about 30 bahaar of cloves, all of which will be properly registered and made to avail ... '.

The Dutch immediately sent out orders to all other townships and villages on the island that they should pledge obedience 'to the King of Ternate and to us on behalf of his Excellence'. They followed this up by making sure that the chiefs came to see them at Taffaso to promise allegiance to both.[41]

A few days later a Dutch force of ten sail, a few fast core-core vessels, 500 Dutchmen and 600 blacks, was reported ready for an attack on the island of Tidore, but the plan was abandoned, and instead the Dutch force sailed away to the western side of the nearby Makian to seize the Portuguese fortification of Taffaso. The Admiral sent messages to the Dutch ships waiting at Tidore, to inform them of the progress and order them to sail south as soon as possible to join him at Makian to reinforce and occupy Taffaso on the western coast. But first they should send two skippers, named as Dirck Allertsz of *Patania* and Cornelis Lurensz of *Duyfken*, to sound the roadstead at Makian and report these depths and anchoring grounds to van Caerden, who would send the advice also to the ships' councils.

Ternate, the Spice Island where *Duyfken* came to an end in 1608. (Seventeenth-century engraving courtesy of Dr N. Sjerp.)

By 27 June 1608, after these orders had been carried out, all the fleet were anchored before Taffaso, except for the ship *Bantam* which had gone to Ternate to help safeguard the new Dutch Fort Oranje at the town of Malayo.

After the ships' arrival at Makian, *Duyfken* was sent away with the military captain (named Schott) of Malayo's Fort Oranje and some local people of Ternate on a reconnaissance to assess that position and also determine what other places on the island would need to be fortified. The plan was to leave Captain Schott at Malayo with his soldiers and also as many locals as were required to bring the garrison total to 100, because Ternate was the richest cloves island of all Maluku, according to Admiral van Caerden.

The situation in the Spice Islands seemed to be developing satisfactorily. But nobody at this time in July 1608 could predict the malevolent unseen workings of nature among these unstable volcanic areas above and below the sea. To quote the bewildered Admiral van Caerden:

> Then are by the Will of God, on the first of Julio the *Walcheren* and *China* shipwrecked and driven ashore against the reef, of which crew, money,

cannon and some rigging and sails are salvaged, including most victuals from the ship *China*; victuals from *Walcheren* were of little value ...

Walcheren carried Vice-Admiral Adriaan Martensz, and *China* was skippered by Kornelis Maartensz; both were apparently young ships, having left home between April and May 1606. The loss was inexplicable to the Dutch:

This accident causing the loss of the ships did not happen during a fearsome or heavy gale, but by the swell and surf of terrific high seas, caused by a storm that blew at sea, or had blown, and all other ships were in great peril ...

There was some muttering about 'an act of God', but the mystery may well be explained today by an underwater earthquake and the resulting upheaval and monstrous waves, or the Japanese term, 'tsuna'mi', for a series of long, high sea waves caused by a disturbance of the ocean floor, or a seismic movement.

The loss of the two ships troubled the admiral, but he recorded the disaster, reorganised the crews of his other ships and continued to salvage as much as was possible. Several weeks were to pass before van Caerden, early in August 1608, referred briefly to another small loss in the southern anchorage beside Ternate's town of Malayo, presumably from the same time when *Walcheren* and *China* had been suddenly swept to their doom on the northern coast of Makian:

... the *Duyfken* has been brought here behind the reef because of her leaking and careened ashore. After much effort and work done, found that the whole of her hull has given in and because of her age [she] cannot last any longer, because since ashore the knees have completely come loose and two beams in the hold have snapped and her sides so much distorted that nothing can be done and she has to be condemned as a wreck ...

In November 1608 the chief merchant at Bantam, Jacques le Hermite, wrote a letter to the VOC directors in Amsterdam, reporting the success of Admiral van Caerden's operations in Maluku. He added a reference to a minor matter:

The *Duyfken* is burst in the doubling [a lining of extra planks below the waterline], so that it was necessary to send some ships thither with food and other things.[42]

It seems most likely that *Duyfken* was irreparably damaged by the same mysterious upheaval of water in early July 1608 that had wrecked the two bigger ships *Walcheren* and *China*.

Marit van Huystee's paper[43] throws more light on the last days of the Little Dove. What was probably the last surviving document written aboard *Duyfken* was the resolution taken by the ship's or the fleet's council on 8 June 1608, one of the signatories being Jan Lodewycksz van Roossengijn. The Dutch had decided to send the ships *Kleine Son* and *Duyfken* to Gammalama on the island of Ternate to see if they could capture a ship that was anchored in the roads, or set in on fire. Cornelis Laurensz and Dirck Allertsz, the captains of *Duyfken* and *Patania*, who were mentioned in Admiral van Caerden's journal, also signed the resolution.

The ship *Patani*, yachts *Duyfken* and *Jager*, the big sloop of the *Walcheren* and the frigate of the ship *Gelderland* were manned by crews of the large ships. They arrived at Makian on 20 June and seized the fortification at Taffaso on the following day.

A message was sent to the larger ships of van Caerden's fleet that they could now sail for Makian, while the captains of *Patani* and *Duyfken* were sent out to sound the roads of Taffaso.[44]

According to van Caerden, on the first day of July 'there was a terrible sea without any wind, therefore the ships could not sail away from the shore', and because of this 'extraordinary weather' the ships *Walcheren* and *China* were driven ashore and wrecked.

Ms van Huystee noted that all the islands in Maluku were in fact tops of volcanic mountains. The different sources were not consistent about the island where *China* and *Walcheren* were lost, but she was convinced from her reading that it was the island of Makian, not Ternate. Captain Saris had confused the situation:

> The first of September, arrived a small Pinnasse of the Flemmings from Mackian by whom we understood of two ships called the *China* and the *Dove* were cast away riding at anchor afore Makian with very little wind at West, which makes such Sea that it is not possible for ships to ride there, by reason it is foule ground, and very deepe water, as seventy and eightie fathomes [about 130–150 metres].

Saris had misunderstood. Possibly he had heard about the loss of *Duyfken* at the same time, and had mixed up different names and events. As a consequence it had been assumed for a long time that *Duyfken* was lost at Makian. The Saris reference to a westerly wind indicated that *Walcheren* and *China* could only have been borne down upon a lee shore at Taffaso on the west side of Makian. The site can be conclusively identified from the actions van Caerden took when he heard about the loss. Both towns mentioned in the following quotation, Taffaso and Noffaquia, are Makian towns and therefore it must have been Makian where the two ships *Walcheren* and *China* were lost, and not Ternate:

After the loss of the ships, I immediately went from the other side of the island to Taffaso ... and went in search for a better roadstead, and found this before a town called Noffaquia [to the north-eastern end of Makian], where to we took the ships.

Paulus van Caerden had left Makian on 20 July and sailed to the town of Malayo on Ternate, where he arrived next day. According to the admiral, much work had to be carried out there, including repairs to *Duyfken*. However, she could not be repaired, although people had obviously tried to save her. She was probably stripped of fittings and broken up, and it seemed unlikely, although not impossible, that any remains of the ship could still be found at Ternate, Ms van Huystee concluded.

In the seventeenth-century records the little ship *Duyfken* represented a relatively small loss, soon to be written off and forgotten. She was probably built in 1599-1600 as the second *Duyfken*, and the VOC had bought her secondhand in 1603 for 2200 guilders. Her faithful service right up to her very last day afloat had repaid the purchase price many times over.

Yet on her wings the modest Little Dove carried the then unrecognised story of the European discovery of Australia. It was a sad ending for a great little ship.

AFTER THE Spanish victory at Ternate in 1606 the Dutch held on at their strategic fort of Malayu and also built Fort Nassau on the northern side of the island of Motir not far away, and then another, Fort Willemstadt on the north-west coast of Ternate. On 23 November 1609 the Dutch defeated the Spanish-Portuguese garrison on the island of Bacan (or Bachan) and renamed the captured fort as Barneveldt. They continued building new forts among the Maluku islands.[45] The Spaniards responded for a while, but within a few years the Dutch were effectively in control at Ternate and elsewhere in Maluku.

Willem Jansz of Amsterdam

ALTHOUGH THE full name of the skipper of the *Duyfken* which came to Australia in 1606 was Willem Janszoon (as with the English Johnson, or son of John), his signature on documents showed him as Willem Jansz. — with a full stop. His parents were Volckert Jansz, who was also on the VOC payroll, and Elisabeth Claessen. He was given the suffix 'from Amsterdam' to distinguish him from other mariners of the same name.[46] The records of his distinguished career as a skipper, admiral and administrator are not extensive, but he emerges as a skilful mariner and decent man. He does not seems not to have been part of the hard-line extremist actions of Jan Pieters Coen, Governor-General of the Indies during his time. Coen's virtual extermination of the people in the Banda Islands to gain control of their cloves production, and the infamous 1623 Amboyna Massacre of British traders made him notorious amongst early Dutch administrators.

Willem Jansz first appears in the records in 1598, in a signed statement relating to the voyage of a merchant fleet to the East Indies under command of Admiral Jacob Cornelis van Neck. He was presumably steersman on board the ship *Hollandia* in this second such Dutch expedition, and the skippers, steersmen and merchants were required to hand over their journals, charts and other documents to either the admiral or the directors of the company, the Old East Indies Company. The VOC, the United East India Company, would not be established until 1602.

In the next year he is on the master roll of *Hollandia* as Willem Jansz from Amsterdam, uppersteersman, and in 1600 he is listed as steersman on *Gouda*, another ship in the same fleet. Thereafter:

1601: Skipper on the ship *Lam*, in a fleet of three ships under command of Joris van Spilbergen.

1603: Willem Jansz is shown as skipper of *Duyfken* among the thirteen ships in Admiral Steven van der Haghen's East Indies fleet which departs from Holland.

1605: Captain Willem Jansz, with Jan Lodewycksz van Roossengijn as supercargo, embarks on *Duyfken*'s voyage of exploration of

'Nova Guinea' which takes them to today's Cape York Peninsula, returning to Banda in 1606.

1606: Willem Jansz given command of the bigger ship *West Friesland*, which departs from Bantam, laden, for home but is wrecked early in 1607 near Mauritius. Jansz and his supercargo, Olpherts, are taken on a yacht — built at Madagascar and so named — to Bantam, where he is found on 24 November by Admiral Cornelis Matelieff with five ships.

1607: On 4 December Jansz leaves Bantam as skipper of the yacht *Madagascar* on a secret mission via Banda to Amboyna to inform Governor Houtman about the expected arrival of two English ships, and a plan to forestall their trading activities. After this successful mission, the yacht is abandoned as a wreck.

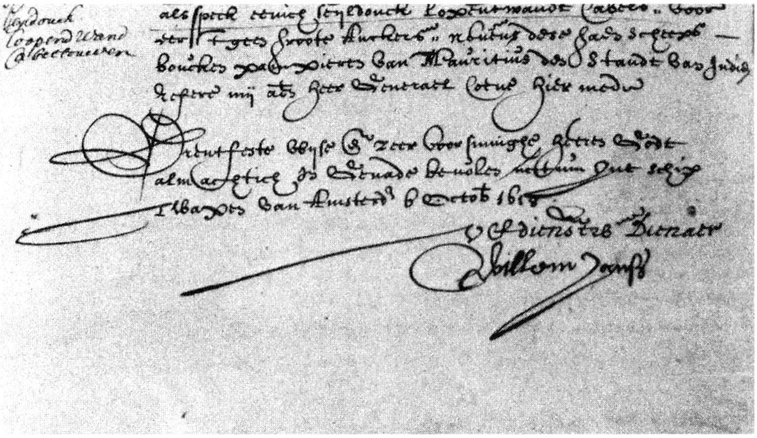
Willem Jansz

Willem Jansz' signature at the foot of a letter to the Company at Amsterdam dated 6 October 1618.

1608: On 20 April, at Amboyna, Jansz becomes skipper of the yacht *Delft* and sails with Admiral Paulus van Caerden to the Island of Ternate, where he is actively involved in the capture of nearby Makian Island. On 3 August, his term of service expired, he departs for home with a letter from van Caerden highly commending him, but at Bantam on the way he is persuaded by Governor L'Hermite to stay and accept a new commission. As both skipper and merchant of the ship *Eendracht* he departs from Bantam on 3 December for Gresik (Java) or Macassar (Sulawesi) for a cargo of rice to be taken to the Maluku islands. (In this year his old ship *Duyfken*, badly damaged, is abandoned at Ternate.)

1611: Jansz ends his second term of service with the VOC, sails back to the Netherlands as upper-merchant on the ship *Oranje*, arriving at Zealand on 23 August. By 1 September he has accepted a new term of service with the VOC. On 29 December he sails on *Oranje* for Bantam.

1613–14: Jansz and *Oranje* are involved in several naval actions against the Portuguese; in one of them *Oranje* is dismasted. The yacht *Halve Maen* brings bad tidings to VOC Governor-General Pieter Both at Bantam about unrest with the local people on Solor, an island near Bali, and Jansz is sent there as acting governor. He manages to re-establish a peaceful coexistence on the island.

1615: Jansz ordered to return to Amboyna. He returns to Zealand the following year.

1617: He accepts another term of service, now as an upper-commissioner and merchant and recommended to the Governor-General of the Indies for a position as vice-governor or naval commander.

1618: Willem Jansz departs Holland as upper-merchant on the ship *Mauritius*. On the way to Bantam, sailing the Indian Ocean southern route, he makes a landfall on 31 July on the Western Australian coast, south of Exmouth, at a place named Willems River, later shown to be Yardie Creek. After arrival at Bantam, he is asked by Governor-General Coen to advise the Council of the Indies, which accepts his advice about enforcing the Dutch presence at Jakarta.

1619: Jansz is involved in Coen's unresolved military dispute with the English rivals at Jakarta. Jansz is appointed as a member of the Council of the Indies. On 23 May he is with Coen's fleet in an operation to lift an English siege of Jakarta. They capture the town and burn it. On 18 September Admiral Jansz takes a fleet of six ships to the western coast of Sumatra and captures four English ships. Governor-General Coen and his Council reward Jansz with a golden chain, valued at 1000 guilders, for his bravery.

1623: In new Jakarta, now renamed as Batavia, Jansz renews his service for three years. He is appointed Governor of Banda, with duties to promote agriculture, establish good foundations for school and church, uphold law and order, and bring order to the administration. He starts a lottery to obtain necessary funds for schools on the Banda Islands, organises a form of leaseholds for land and encourages trade with the Kei and Aru Islands, returning to Batavia in 1627. Then he commands a fleet which undertakes operations along the coasts of India and Persia.

1628: Back in Batavia again, pending return to the Netherlands, Jansz takes an advisory role in the defence of Batavia against an attack by the nearby Kingdom of Mataram.

1629: Home at last. He reports to the Dutch States-General and the Prince. This ends the career of Willem Jansz. He is believed to have died in 1638. Little is known of his private life, and apparently he did not marry.

According to written references to Jansz from the directors of the Dutch East India Company, he was 'a very able and straightforward man who gave us much satisfaction with his report on the East Indies trade ... it were wished that the Company had many such servants in India'. Governor-General Pieter Both wrote of him: 'Willem Jansz is of a natural good nature, modest and pious ... and has not had the least of problems with his crew and always observed good order on his ship'.

Sadly, he lived and died without knowing he had discovered the Sixth Continent. Captain James Cook, an Englishman of equal stature, would only discover the eastern coast 164 years after Willem Jansz had found and charted Cape York Peninsula — the Australian coast of European discovery.

The Foundation of a Dream

THE *DUYFKEN* 1606 Replica Project was born on 5 February 1994 at the home of Dr Cornelis de Heer in Applecross, Western Australia, beside a 1:31 scale model of the ship he had been working on for about five years, a beautiful and intricate work of art about one metre long. Dr de Heer, Michael Young and I formed the committee to plan for the raising of public support and funds to build an authentic replica of the ship at Fremantle.

Michael Young and his parents, with the common Dutch surname Jong, had migrated from the Netherlands after the Second World War, but at primary school his teachers decided his name should be pronounced in the proper Australian way, as 'Young', and then altered the spelling as well. Michael accepted this, but at university an interest in history led him to study his family tree, where he found several members with the Dutch name Jansz. He liked to fantasise that he was descended from Willem Jansz, skipper of *Duyfken*. His mother had been a de Ruyter, the name of the Netherlands' most famous naval hero, seventeenth-century Admiral Michael Adriaansz de Ruyter.

Young moved to Western Australia in 1987. Late in 1993, after the replica of Captain James Cook's *Endeavour* was launched at Fremantle, he wrote a letter to the *Fremantle Herald*, urging that a *Duyfken* replica be built at Fremantle by the *Endeavour* team, and calling for contacts from interested people. Early the next year he wrote similarly to the *West Australian* as *Endeavour* was nearing completion and scheduled to sail the following year for its home port of Sydney, a disappointment for Western Australian supporters of that project. Many people signified interest in *Duyfken*.

At the meeting at de Heer's home, we agreed Young would be the secretary of the replica project and de Heer would complete his model and be the historian and naval architect. Young and I would together publicise the replica idea, and organise public support. To attract media attention, I was to organise a '*Duyfken* 1606 Rediscovery Expedition'.

From that day, Young applied himself with feverish energy to the cause of a *Duyfken* replica, and his relentless determination would make our visionary project idea unstoppable.

The wall of Dr de Heer's study was dominated by a dramatic Dutch engraving, about two metres wide, of Amsterdam harbour in 1606, crowded with about sixty sailing ships of various sizes. 'Most of these ships would have been built between 1604 and 1606', he told us. 'They must have been a reflection of Dutch ships at the time when *Duyfken* came to Australia in 1606.'

He motioned us closer to the study wall to examine a smaller ship, which appeared to be of the same design as the *Duyfken* model he was building. He told us the sixteenth- and seventeenth-century Dutch shipbuilders produced their many hundreds of ships without formal plans. 'But there's pictorial evidence of what those ships looked like. Here's some of it from the very time and place we're interested in.'

Michael Young

Most of the ships in the picture were square-rigged and three-masted, with the mizzen mast carrying a lateen rig — a triangular sail on a long yard at an angle of about forty-five degrees to the mast.

He explained that he had been working from plans in a twentieth-century Dutch book he bought in Java after the Second World War, dealing with a historic seventeenth-century ship, *Halve Maen* (Half Moon), built in 1608 and similar to *Duyfken*. She was of the same general dimensions as *Duyfken*— sixty-three Amsterdam voet long (the Amsterdam voet, or foot, was the equivalent of eleven inches or 0.283 metre) and seventeen voet three duim (inches) wide, with a cargo capacity of about sixty tonnes and crew of twenty.

Halve Maen was sent south-west across the Atlantic to America where the English Captain Henry Hudson (1550–1611), in service of the VOC, commanded her in a search for a north-west passage to the East Indies and on a commission to explore North America. He discovered Hudson Bay in 1609 but was said to have disappeared a couple of years later, after being cast adrift by a mutinous crew.[47]

In 1909 the Netherlands Government presented New York City with the an accurate *Halve Maen* replica to celebrate the 300th anniversary of her exploration of the Hudson River. *Halve Maen* was photographed then moored beside a wharf at New

York, but evidently she was neglected by the Americans after the ceremony, and within a few years she had either rotted to destruction or been burnt.

Duyfken, as we believed at this time, had been built in 1594–95, fourteen years before *Halve Maen*, but de Heer concluded that regular refits and modifications after her return from voyages to the East Indies would have seen the *Duyfken* closely resembling the slightly later *Halve Maen*.

De Heer had earlier built a scale model of the Dutch ship *Zuytdorp*, wrecked off the Western Australia coast in 1712, which was purchased by the Western Australian Maritime Museum. His scale model of *Duyfken* became the valuable visual focus for the replica concept.

We decided at this early stage that the replica must remain a Western Australian ship, home-ported at Fremantle, thus attracting support from people who had lamented the loss of *Endeavour* when she departed for her home port of Sydney. The *Endeavour* replica had cost about $16 million and had taken about six years of starting and stopping, because funds kept running out, so our much smaller ship would be a speedier undertaking and cost a good deal less.

Young brought the *Endeavour* Replica Foundation's chief executive John Longley, its master shipwright Bill Leonard, and the Netherlands Vice-Consul in Western Australia, Jan Gunnink, to inspect Dr de Heer's *Duyfken* model, while we explained our replica project to them. They were greatly impressed by the model and our idea. We wanted Longley to become head of the *Duyfken*

Engraving by Willem Jansonius of ships in Amsterdam harbour, 1606. (Courtesy Dr C. de Heer.)

project, but after the years of stress involved in the *Endeavour* project, he was not looking for any more. After consideration, however, he was prepared to be listed as patron of the project. Eventually Bill Leonard would agree to head the team to build our ship, but we had a lot to do before that could happen.

Gunnink introduced us to David van Itterson, the Netherlands Consul-General, based in Melbourne, who was to provide regular advice and support, including a grant for translating some *Duyfken* documentary material sent from the Netherlands.

A steady stream of visitors came to admire the model as de Heer continued meticulously to construct the hundreds of tiny parts and assemble them. People marvelled that such a small ship could have made voyages from Holland to the East Indies and then to Australia. Young and I set to work on research and generating publicity in newspapers, magazines, on television and radio across Australia, in England and Europe.

As our research efforts progressed we realised that the already published information about the 1606 *Duyfken* and the circumstances of its inadvertent discovery of Australia was the tip of an iceberg. A dramatic record of four centuries lay buried in old documents held mostly in Dutch archives and museums — accounts of explorations, discoveries, and nations in violent conflict over the fabulous Spice Islands. These documents described the grand events, bloody naval battles, the careers of princes, priests, politicians, admirals, merchants, administrators harsh or compassionate; sometimes little incidents too, and ships and individuals, all waiting to be brought back to life.

Most of this story needed to be decoded from countless pages of baffling old Dutch handwriting and phraseology. Old seventeenth-century English was bad enough, as I had found from earlier book research; a special type of supporter would be needed to cope with this problem. As it happened, Adriaan de Jong, a man of many talents — translator, researcher, artist and draftsman — would find us.

Cornelis de Heer working on his scale model of *Duyfken*, and (right) the completed model.

Adriaan had come to Australia in 1982 with a diploma of Applied Sciences and Chemistry. He had a lifetime interest in historic sailing ships and soon revealed surprising artistic and scientific skills and an impressive talent in translating ancient documents. As a small boy living near Utrecht in Holland he had made his first model sailing boat from a worn wooden clog. It was not very successful when he launched it in a canal which ran through his historic area of Linschoten Landgoed (estate) with its church buildings dating back to the twelfth century.

In 1994 a supporter of the *Duyfken* project asked Adriaan if he knew anybody who might make some drawings of the vessel for promotional purposes. He offered his services, and after visiting Dr de Heer's model he began to make lively pictures of the little ship. By this time he was intrigued by the *Duyfken* story.

De Jong's background interests served him well, not only in deciphering the various ships' logs and associated documents but also in developing the plans for the construction of the *Duyfken* replica and even pictures of how our proposed old Dutch shipyard might look. He found that the Little Dove had earned surprisingly frequent references in the shipping records, with indications that she was a fast, manoeuvrable and useful ship.

Later, when he managed to translate the formidable journal of the flagship *Gelderland*, with which *Duyfken* had sailed to the East Indies in 1601, Dr de Heer, a skilled linguist, was amazed by his ability to cope with what seemed an impossible task of deciphering the old-style handwriting.

Michael Young was contacted early in 1994 by Ms Ninette Radke, offering support from the Netherlands. He asked her to set up a support group there and also to forward a submission to the Dutch Fugro group, which seemed genuinely interested in early Dutch-Australian maritime history. Already, through their Perth-based Australian subsidiary Fugro Survey Pty Ltd, this firm was sponsoring research at the Western Australian Maritime Museum at Fremantle relating to the *Batavia* shipwreck.

In Holland, a senior Fugro staff member, Huych Mak van Waay, and another history enthusiast, Philip Lach de Bere, became involved in research activities. They contacted the maritime archaeologist Dr Robert Parthesius, and ship model maker Ab Hoving of the Rijksmuseum, a world authority on sixteenth- and seventeenth-century sailing ships, and gathered records on ships of the early Dutch expeditions to the Spice Islands.

Young maintained constant contact with Dutch and Indonesian consular offices, and this was to yield benefits large and small for the project. Together we called on various State Government officers concerned with tourism, and sought their interest and assistance. The committee and supporters, including a good number of expatriate Dutch people, grew in numbers and enthusiasm, meeting at the Young home in East Fremantle, where we discussed what we had learnt of the ship's history.

An Adelaide researcher sent Young a copy of the fifty-five-page booklet on *Duyfken*'s discovery of Australia, that journalist-politician-historian Thomas Mutch of Sydney had privately produced from his 1942 Royal Australian Historical Society paper. This pioneering work presented a remarkably clear and accurate account of the relevant events — as we were to confirm from our own research activities — and it included a copy of the seventeenth-century discovery chart of the skipper of *Duyfken*, Willem Jansz.[48]

Mutch had searched in vain in Dutch and Australian histories for a connected narrative of the voyage. The log of *Duyfken* had been lost; like many other documents of historical importance, its value unrecognised by those who handled it in those busy and exciting years of the Dutch East India Company. Mutch speculated that if the log had been forwarded to the headquarters of the Company in Amsterdam, it might have been sent on to the Company's cartographers for the preparation of charts — and not returned.

Mutch did not exclude the possibility that the log was stolen, but whatever happened, it had disappeared. There was some evidence, he noted, that it was sought in the Indies in 1616 without success, and it was certain that some particulars of the voyage and a copy of the chart, were available to Jan Carstensz who followed *Duyfken*'s course to the Gulf of Carpentaria in command of the *Pera* and *Arnhem* expedition in 1623.

Duyfken's 1605–06 chart however had a better fate than the log or journal, Mutch wrote. It was still in existence in Amsterdam when Hessel Gerritsz made his map of the Pacific in 1622 and placed the *Duyfken* geography on it, and also about 1670 when a copy was made which eventually went to the Imperial Library in Vienna.

In 1994 I visited the Mitchell Library to look for the *Duyfken* chart of 1605–06 that had been found by Dr F. C. Wieder in the 1930s and acquired through the efforts of Thomas Mutch. After my bona fides were established, the chart was carefully brought from its secure vault. It was about a metre wide, with its black pen outlines on the faintly sepia surface telling the story of the first European discovery of Australia. To me, this precious chart of Australia, drawn in the seventeenth century, was surely the Australian nation's earliest and most important European historical exhibit.

The Mitchell Library later sent me photographs of the chart and some associated illustrations of islands of the East Indies. Now I was committed to a rigorous search for all available information of the beginning — and perhaps the mysterious end — of the historic vessel.

Adriaan de Jong translated the inscriptions and explained to me that the *Duyfken* chart is actually a compilation of the results of three exploratory voyages, two of which involved *Duyfken* in her one voyage to Australia. It is not known whether Jasper Janssen de Jong was on *Duyfken* or some other ship when he charted the section named 'Nieuw Zelant with the Gounongapi'. The chart appeared to be a seventeenth-century copy of an earlier one, perhaps of the 1606 original. De Jong would have expected the original to have been a proper portolano chart of sailing directions showing the radiating sixteen or thirty-two compass rose lines, but these lines were missing, although there was a compass rose.

De Jong also noted that the inscription on the *Duyfken* chart was in handwriting remarkably similar to that found on two charts of Carstensz. If they were indeed from the same hand, then perhaps a set of charts was made by an artist some time after both the voyages of *Duyfken* and also of Carstensz, a task ordered by the Dutch East India Company to commemorate the events in its corporate history.

Crown Prince Willem-Alexander of the Netherlands made an underwater visit to the 1629 wrecksite of the Dutch ship *Batavia* in the Abrolhos Islands, after seeing the excavated timbers at the Western Australian Maritime Museum. The skillful photo by Museum photographer Patrick Baker shows the Prince (centre) patting an anchor fluke, discoverer Max Kramer of Geraldton (left), and Dr Michael Kailis of the *Duyfken* Foundation (right).

Why were the two main explorers, Jansz and Roossengijn, not listed in the text of the *Duyfken* chart? Perhaps the original chart was made or ordered by Jansz, who then gave credit to his two officers Abraham Francken and Jasper de Jonge for their work done on earlier voyages. He had no need to put his own name on his work because he would personally present it to his superiors, who would grant him due credit. Some years later the official copyist would make a copy or copies, by then unaware of who had completed the original.

Just as I had been thrilled by my first sight of *Duyfken*'s 1605–06 chart in the Mitchell Library, Michael Young was excited by a book, first published in 1610, that he bought in April 1994 when he visited the Portuguese Maritime Museum and Centre for Maritime Studies in Macao. This dealt with the career of a Portuguese admiral who had been ousted by the much smaller Dutch fleet of Admiral Wolphert Harmensz in the celebrated Battle of Bantam in 1601 and was involved in another battle with the Dutch later. The book opened with a dramatic picture of *Duyfken* in the middle of this naval battle; virtually a moving picture from nearly four centuries ago of the ship we wanted to recreate.

BY JUNE 1994 our *Duyfken* project had an enthusiastic and growing membership roll of 100. The first of many leaflets, brochures and *Duyfken* Replica kits I produced in conjunction with Michael Young for our supporters, the media and general public, began:

> This message of the *Duyfken* Replica Project Committee is intended to stir Australians now, approaching the year 2001, to be more aware of their earliest known national maritime history. With the story of the tiny Dutch vessel we are raising the flag, signalling new interest in the neglected, little known but entertaining chapter which belongs to the Little Dove...

The indefatigable Young transformed a bedroom in his home into the *Duyfken* Replica Project's first office, and his telephone and fax were alive far into the night recruiting members and opening up channels of information from near and far, especially the Netherlands.

On an expedition to Indonesia Young met my son Graeme, maritime archaeologist and director of the Western Australian Maritime Museum, who was holidaying there. They went together to the island of Ternate, the ancient centre of the Spice Islands, and beside the remains of the Dutch Fort Oranje, Graeme dived and swam about the area where documentary evidence indicated our *Duyfken* had been careened and then abandoned. Water visibility was poor and it was impossible in the time available to decide whether anything of the vessel might be found here. To complicate the situation, Ternate had suffered a major volcanic eruption in 1608; the ship's remains might well have been covered by many tonnes of ash.

At home the replica project committee continued its growth in members and supporters. An early volunteer to the committee was Peter Becu, the manager of information systems for Robe River Mining Co., who offered to set up a list of members and supporters, and prepared the project's first business plan. A former colleague of Young, environmental consultant Charlie Welker, joined in with offers of his services, a West Perth public accountant, Rinze Brandsma, helped set up a constitution, and the Perth legal firm Kott Gunning freely advised on establishing a foundation. Graeme Cocks, a public relations consultant associated with the *Endeavour* Replica Foundation, offered to help and attended meetings.

We found that we had to compete with well-intentioned rivals lobbying for other historic ships, so on behalf of the *Duyfken* committee I wrote to the *West Australian*, urging the various groups to realise that the *Duyfken* Replica Project for Fremantle was already up and running, that they should support it because it would help to establish Fremantle as a historic replica construction centre, and their projects could more easily follow it.

Later in 1994 we had to face the fact that winning public recognition and gathering an enthusiastic band of supporters was not itself going to get the *Duyfken* replica built. We had to 'go with the strength' — persuade a powerful person in the business community to become chairman of a foundation, with a board of appropriate members to show our capacity to do the job and the reputable substance to attract the necessary funds from governments and corporations.

John Longley advised how to go about setting up a foundation similar to the *Endeavour* Foundation, and Charlie Welker began work on this task. Longley also suggested that we should invite the well-known successful Fremantle fishing and pearling identity Michael G. Kailis to be chairman of our foundation.

Unfamiliar with the process of inviting powerful captains of industry to head boards of directors to raise millions of dollars, Young and I approached the matter with due care, and discreetly sought guidance in Fremantle. This led us eventually to an appointment with Michael Kailis' engineering manager Terry Hewitt in the M. G. Kailis Group offices overlooking the extensive marine engineering yard and fleet moorings in the fishing boat harbour. The chief was away for a few days, but Mr Hewitt was encouraging: 'A replica only twenty metres long? See that wharf space down there? We could build it just there, no bother. But I'll pass on your material to Mr Kailis and tell him what you said, and we'll let you know if he wants to talk about it'.

In due course we met Kailis and formally invited him to be our foundation chairman. The challenge was for him and the M. G. Kailis Group, as reputable shipbuilders, to be involved in a project to build a ship of historic importance to Australia. A point of the discussion was Michael Kailis' link with Greece and his own personal link with the sea. The M. G. Kailis Group is the largest privately owned marine corporation in Australia, involved in pearling, jewellery manufacture, prawning (catching and processing), lobster processing, tuna farming, mining, engineering as well as trading on an international basis in many world markets. But it is not a professional fundraising group.

Young and I found Kailis friendly and interested, already generally aware of the potentials and problems of the replica project. At the close of our meeting he told us he would let us know if he was prepared to take up this challenge. We were to learn about three years later that just after we left, Kailis asked an assistant, Kathryn Cartwright: 'What do you reckon?', and she replied: 'Well, it's an Australian project and it's right here; I'd support it'.

A week or so later we got our positive answer. We were with the strength now: Kailis had a reputation for getting things moving and completed. He began the procedure of setting up an interim *Duyfken* 1606 Replica Foundation board.

The initial board members were Michael Kailis, builder Syd Corser, Rinze Brandsma, accountant and yachtsman Geoff Totterdell, Peter Becu, Charlie Welker, newspaper executive Sir James Cruthers, secretary Gloria Jackson — and Michael Young and me. The third member of our original *Duyfken* committee, Dr de Heer, declined because he did not think his health would permit him to regularly attend board meetings, and he was listed as Historic Ships Adviser.

The interim board decided to set up a 'Friends of the *Duyfken*' group to include our current project committee members and supporters. Ms Jackson, who had been an early secretary of the project, resigned from the Foundation board and Rod Murfitt, who was chairman of the Friends, became a member. Later, when he resigned from chair of the Friends, he would be replaced on the board by the incoming Friends chairman, Con Smit.

The Foundation had inherited grand ambitions from the *Duyfken* Project group, but no money; the committee with its hard-working volunteers had raised only enough to keep itself alive. But Michael Kailis and Syd Corser made generous cash donations to the bank account to cope with costs of building the structure of the Foundation.

After the establishment of the *Duyfken* 1606 Replica Foundation a number of committees were set up to deal with various issues that had to be addressed before it could seriously look for funding: a technical committee to study the design of the ship, its cost and end use; a committee to choose the site for a shipyard; one to design the shipyard and another to study the means of raising the necessary funds.

Graeme Henderson agreed to join the board. It was obvious that the involvement of the Maritime Museum in the design and construction of the ship, would enhance the public profile of the project. The Museum director pointed out that this would require the approval of the board of the Maritime Museum, and then any proposed buildings had to be accepted by the City of Fremantle and the appropriate State Government authorities. His board would want to have the replica built as an authentic, sailing reconstruction of the original 1606 *Duyfken*.

One option for the shipbuilding site was the Fremantle waterfront building which had been used for *Endeavour*, but this was on land whose lease was limited and would be too expensive. An area of Fremantle Harbour Trust land was offered for lease, but was also rejected. A car park area of the Maritime Museum in Cliff Street, Fremantle, appeared the most attractive prospect for the Foundation, but there were some difficulties about the existing use of this section, and in 1996 the adjacent lawn-covered area on the southern side of the Museum was selected.

The question of an engine became an early issue for the Foundation board. Initially the chairman and most members of the board regarded an engine as a practicable necessity for the replica. Kailis, who had built more than 100 sturdy steel 'K' boats (mostly bigger than the wooden *Duyfken*) for his fishing and pearling fleets, was an experienced and strong advocate of an engine, for safety reasons and for manoeuvring the ship in and out of harbour in contrary winds. And it occurred to some members of the board that the replica might not be insurable nor allowed out to sea without an engine — it might not meet the marine regulation standards required for the crew and any passengers.

This debate continued while the many other problems were being addressed, such as the overall cost of building the ship, raising the necessary funds, public promotion of the project, the design (considering that no plans were available because the Dutch builders of *Duyfken*'s time didn't use plans), the timber, the basic method of building, the end use.

Maritime archaeologists within the Maritime Museum were firmly opposed to the idea of an engine. Nick Burningham, the traditional watercraft specialist with the Museum who had been appointed as the *Duyfken* replica's nautical archaeologist-designer, was asked to report on the issue, and his resulting paper did not favour an engine. After a few months an increasing number of Foundation board members felt the same, and realised that this small replica ship would have little room for paying passengers, for whom the regulations might apply. It might sometimes require an escort vessel which could well carry paying passengers.

Charlie Welker developed a business plan and costed the replica at $3.5 million — fortunately nowhere near the cost of the *Endeavour* replica. Bill Leonard, the master shipwright who had built *Endeavour*, made an independent study and came up with the same figure.

A question of some concern was the shrinking of the timbers of the replica if it were to be exposed to drying winds and sun over several years. At the time when the original *Duyfken* was built in Holland the hull would have been completed and launched in about six months, so the seasonal change then was no worry, but in windy Fremantle the *Endeavour* Replica — even in an enclosed shed — had suffered problems of this nature over its six years of construction. The *Duyfken* replica would need a shed to protect it from the elements and for security of the ship and tools and equipment, but a permanent building would not be acceptable either to the Museum or the Fremantle Council.

A memorandum of agreement was drawn up between the Foundation and the Maritime Museum, concerning terms and conditions for the use of the Museum land and the involvement of the Museum in the project, in which the shipyard shed was to be temporary. Eventually all parties concerned agreed to the document.

In December 1994 the Foundation board members approved my plan for a research expedition to Cape York Peninsula to determine the likely places where *Duyfken*'s boat crew would have made landings, and to look for the Aboriginal elders' version of the events of 1606. This was intended to be followed eventually by a '*Duyfken* 1606 Rediscovery Voyage' — to sail the *Duyfken*'s discovery course from Banda to Cape York Peninsula. It should help publicise the replica project and attract funding, I asserted. But the Foundation had no funds for such activities, and I saw that any donations in the near future would be directly applied to getting the ship built. It was left up to me to organise and fund the expedition, if I could.

After a couple of promising leads, each of which came to naught, Michael Young suggested that the Fugro Group, through its Australian subsidiary Fugro Survey Pty Ltd, might be prepared to sponsor my research expedition. He took me to meet the marine manager in Perth, Grey Roughan, who was already interested in the replica project.

Fugro Survey and its associated company, Omnistar, are both Australian companies, with Australian directors resident in Australia and some ownership in the Netherlands. Fugro specialises worldwide in leading-edge computer and satellite technology for surveying and navigation on, over and under land, at sea and undersea. The company agreed in principle to support my planned first *Duyfken* research expedition, and Mr Roughan invited me to submit a budget for the venture.

Meanwhile, in February 1995 — only twelve months after our project was born — the Western Australian Minister for Resources Development, Colin Barnett, officially launched the *Duyfken* 1606 Replica Foundation at a large function in the *Batavia* Gallery of the Maritime Museum in Fremantle.

LATER IN THE YEAR Marit van Huystee, a historical research worker at the Maritime Museum, produced her paper for the Maritime Archaeology Department entitled 'On the Yacht *Duyfken* (1601): The first European ship to explore the Australian coast'. This was the paper that revealed to us all that the *Duyfken* that explored the coast of Australia in 1606 was not the same vessel that had been launched in 1595, and had accompanied the first Dutch fleet to the Spice Islands.

The *Duyfken* of the so-called First Fleet of the Dutch that sailed in 1595 for the Indies, she found, was built in May 1594 in Amsterdam at a shipyard called Uylenburg. No building plans or other specifics of the small ship were found.

The *Duyfken* was then renamed as *Overijssel* (after a Dutch province) and sailed for the Indies in May 1598, arriving there in November. In January 1599 she went back to the Netherlands, arriving at Texel in July of that year.

Late in December 1599 *Duyfken/Overijssel* sailed for the Indies for the third time (once as *Duyfken*, and twice as *Overijssel*), arriving at Bantam in September 1599. She sailed back to the Netherlands, arriving there in June 1602.

But Ms van Huystee found that another ship called *Duyfken* — the *Duyfken* which was to sail to Australia in 1606 — departed Texel for the Indies on 23 April 1601, so the *Duyfken/Overijssel* built in 1594 and the *Duyfken* which sailed in 1601 could not be the same ship. Her table shows the departure and arrival dates of the ships involved:[49]

Ship/Tonnage	**Departure Netherlands**	**Arrival Indies**	**Departure Indies**	**Arrival Netherlands**
Duyfken 50 tons Pinas 1594 Amsterdam Uylenburg	2-4-1595 Capt. Simon L. Mau Texel (1st Fleet)	6-6-1596 Engano	25-2-1597 Capt. Cornelis Adriaansz Bali	11-8-1597 Texel
Overijssel 50 tons Pinas 1594 Amsterdam Uylenburg	1-5-1598 Capt. Simon Jansz Hoen Texel	26-11-1598 Bantam	12-1-1599 Capt. Simon Jansz Hoen Bantam	19-7-1599 Texel
Overijssel 50 tons Pinas 1594 Amsterdam Uylenburg	21-12-1599 Capt Thijmen Michielsz Texel	1-9-1600 Bantam	9-9-1601 Capt. Thijmen Michielsz Bantam	June 1602 Texel
Duyfken 60 tons Jacht Amsterdam	23-4-1601 Capt. Willem Cornelisz Schouten Texel	26-12-1601 Bantam	25-8-1602 Capt. Willem Cornelisz Schouten Bantam	17-2-1603 Vlissingen
Duyfken 60 tons Jacht Amsterdam	18-12-1603 Capt. Willem Jansz Texel	11-12-1604 Bantam		

In March 1998 Michael Young made an expedition to Indonesia to search for remains of seventeenth-century Spanish and Portuguese forts. With the aid of a seventeenth-century Dutch chart passed on to him by Perth historian Rykus Kaijer, Young found Duyfken Island, in the Bay of Batavia about 5.8 kilometres north-west of Jakarta. Kaijer had found the excellent engraved chart during a recent visit to Holland, in a book entitled *Jaan Nieuhofs zee en landreize door verscheidene gewesten van Oostindien*, published in Amsterdam by Jacob van Meurs in 1682. Young discovered that the historically named island, only thirty metres by 100 metres, had been subsequently renamed as Pulau Dapur (Kitchen Island) because fishermen landed there to cook their catch. To go ashore and explore the sandy little islet, Young had to swim in from a dinghy. Locals told him that a cyclone in the 1950s had reduced its size, but it was now being re-established.

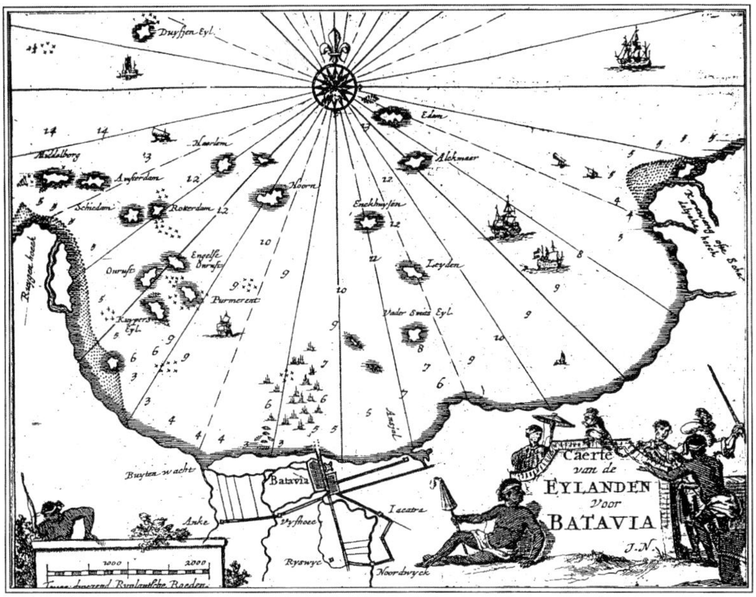

Duyfken Island (top left) in a seventeenth-century chart of the Bay of Batavia. With this chart, from Perth historian Rykus Kaijer, Michael Young found and visited the little island in March 1998.

Young also went to Makian Island in the Maluku Spice Islands where, with the help of villagers and more historic documents from the Netherlands he found the remains of Fort Taffaso, the Spanish fort which the *Duyfken* crew helped capture for the Dutch in 1608. Later Young and another Dutch specialist Jaap Keppel found an unrecorded Spanish fort on nearby Tidore Island which had been missed by previous expeditions. The *Duyfken* was believed to have been involved in raids on a fort there.

IN OCTOBER 1998 the *Duyfken* 1606 Replica Foundation answered a challenge from Shark Bay Shire Council president Les Moss, who claimed that the first Dutch landing on Australian soil was on Dirk Hartog Island, Western Australia, in 1616.

Mr Moss asserted that the inscribed pewter plate erected by Dirk Hartog, of the ship *Eendracht*, on this island off the Western Australian coast was the first European landing on Australia — and that there was no evidence of any landing by the crew of *Duyfken* at Cape York Peninsula in 1606.

We replied, on ABC Radio and in our own *Duyfken Report*, that not only was there quite convincing evidence that *Duyfken* was the European discoverer of Australia, as shown by the surviving 1606 chart, but it could not seriously be denied that sailors from *Duyfken*'s boat came ashore during their survey.

The journal of Jan Carstensz, who commanded the next Dutch expedition to Cape York Peninsula in 1623, referred to the Batavia (now called Wenlock) River where a sailor of the *Duyfken* was speared by Aborigines in 1606. Then, on 15 May 1623, Carstensz made specific reference to 'those of the *Duyfken* who landed here'.

I recalled my visit in 1995 to the first Dutch landfall shown on the *Duyfken* chart, the Pennefather River, with its note: 'River with the Bush', indicating that they had come ashore there, and also the competent surveys far up other rivers of the peninsula that I had observed in my exploration. The Carstensz journal also referred to frequent landings to search for water, and there could be no doubt that the smaller ship *Duyfken*'s crew necessarily did the same.

Marit van Huystee said evidence for *Duyfken*'s crew having landed could be found in the archived records of the Dutch East India Company, in letters, memoranda and journals written by sailors and officials who had read *Duyfken*'s logbook. She referred to instructions written by the directors of the Company which specifically mentioned voyages to the South Land to find what might be there for the Company's benefit. 'How was it possible to find out about trade possibilities without going ashore?'

Research coordinator Nick Burningham added that the first time a factually observed part of Australia's coast was incorporated on a published World Map was the addition of *Duyfken*'s discoveries to a map produced by Hessel Gerritsz in 1621, while Hartog's discoveries were first shown on a globe of 1625 and a map of 1626.

None of this diminished the importance of Dirk Hartog's voyage, he said. The *Duyfken* Foundation was committed to increasing awareness of the role of Dutch navigators in Australia's history: 'We wish Shark Bay Shire every success in their celebration of those histories'.[50]

THE TOUGHEST JOB in the replica project had to be the continual challenge of the Foundation's quiet achievers, its fundraising committee.

Master shipwright Bill Leonard and his team could hardly fail to sense their accomplishment and the justified reputation as the world's most skillful historic shipbuilders today; the design team could be proud of the results; the *Duyfken* was a story of a lifetime, and it seemed everybody was learning to love the Little Dove.

But it all had to be paid for. Away from the exciting activity of the shipyard with its fire-bending smoke, the busy sawing, planing, chiselling, whirring of drills and hammering of trunnels, was the worrying work of finding the money to keep it all going until the launching, and beyond.

The fundraising committee consisted of the Foundation chairman Michael Kailis and board members Sir James Cruthers and Syd Corser, with public relations consultant Graeme Cocks. Foundation deputy chairman Rinze Brandsma, who had lent support with Peter Becu in the early formation of the project, joined the fundraising planning and operations when it was appropriate.

Michael George Kailis, CBE, AM, CitWA, Hon. DSc (Curtin University) — internationally recognised for his successful business activities which included building the largest privately owned marine corporation in Australia, the M. G. Kailis Group of Companies — had a significantly humble beginning.

His crowded *curriculum vitae* shows that he was born in Perth in 1929 to parents who migrated from the Greek island of Kastellorizon, off the southern coast of Turkey, where his grandfather owned a sailing ship. Michael grew up in Perth, learning to speak English when he attended a kindergarten in Pier Street and the suburban Highgate Primary School. Later his family moved to Darwin, where he spent several years until he was evacuated on the MV *Koolinda* to Perth in 1942 after Japanese air attacks.

Aged thirteen he had to work to support his family, but a few years later, encouraged by his cousins, he returned to part-time studies at Perth Boys (State) High School while working full-time. In one year he gained full matriculation to The University of Western Australia. He commenced studies in geology and distinguished himself playing rugby, and after a year at UWA he moved east to the town of Broken Hill in NSW, where he undertook a course of mine management. He subsequently returned to Western

Australia and began an outstanding career of achievement in the marine and fishing industries in Australia and internationally.

Another driving force was newspaper executive Sir James Cruthers. Over a couple of years of development of the project and construction of the ship, he was unceasingly dedicated to the cause, with a long-term view and not afraid to ask people to help. Week after week he ploughed through long lists of potential individuals and corporations for support, contacted them or consulted other members of the fundraising section to organise meetings, where possible accompanied by the Foundation chairman.

A tireless fundraiser of the *Duyfken* Foundation, Sir James Cruthers, at the replica.

He was familiar with the subject of Dutch maritime exploration. As a Perth journalist he had taken part in a West Australian Newspapers Ltd – *Daily News* expedition in 1954 with Phillip Playford to investigate the *Zuytdorp* wreck, near Kalbarri.

During the Second World War Cruthers was a pilot in the RAAF. I had first met him before that when we were both sixteen-year-old juniors at the *Daily News*, where we both worked as journalists after the war. He was a modest, easy-going personality, and a very serious professional. He left West Australian Newspapers Ltd to set up and run TVW Channel Seven, the State's first commercial television. In recent years he has been chairman of the *Sunday Times* section of Rupert Murdoch's Newscorp.

The process of organising and fundraising was to take longer than the building of the ship. Building one from so far back in time was such an abstract concept to promote, Mr Cocks found. People asked: 'Why build an unknown ship?'. His answer was that it was precisely because the *Duyfken* story was so little known — despite a couple of years of media interest generated by Messrs Henderson and Young. 'Unlike the *Endeavour* project of building a famous ship to tell a well-known story, we were to build the ship to make the story famous. But the Foundation had no track record to prove to potential sponsors that we could complete the task.'

Cocks remembered: 'Syd Corser came up with a good idea: he asked several of his business colleagues to donate up to ten hours of their time to come up with a plan to raise the funds. As a result Kerry Stokes, Denis Horgan and others came into the picture.

They met on a couple of occasions and helped form a strategy, the first part of which was to gain the support of the Western Australia Government, because without this backing the whole project had no chance of success.

'So Michael Kailis and Kerry Stokes arranged to call on the Premier. Kailis drove his Mercedes to West Perth to pick up Stokes and took him down to the Capita Centre in St George's Terrace, central Perth. He parked and reached in his pockets for coins for the parking meter. Not a cent, so: "Have you got a couple of dollars on you, Kerry?". But these two of the State's most successful businessmen didn't have a dollar between them. They were in luck that day however; the Mercedes escaped the attention of the parking inspectors.

'And some weeks later, after a great deal of discussions with the State Treasury, the Premier eventually agreed to provide $500,000 to support the project. *Duyfken* was under way! We could hire a chief executive officer.

'Since the project began we have produced dozens of sponsorship proposals. We have had our greatest support from governments, although we do have a good range of corporate supporters. The Dutch Ambassador helped us secure the support of the Dutch and Australian Governments. We have not yet [as at June 1998] secured a corporate naming rights sponsor', Cocks concluded.

The Foundation board laid down the policy for the raising of funds, the design and construction of the *Duyfken* and its end use. Noel Robins, a top-ranking Australian yachtsman, was appointed in 1996 as executive director, and it was his responsibility to put everything together and make it work.

Media reports of the man who skippered *Australia* in the 1977 America's Cup Challenge presented him as 'Mr Nice Guy of Yachting' in contrast with 'some egotistical millionaires and assorted loudmouth larrikins'. To quote *Woman's Day*, 3 October 1977: 'It is refreshing to find someone so talented and so totally unassuming, so naturally just a plain human being as Noel Robins. He was without question the quietest man in Newport'. But the writer went on: 'At sea he is the total competitor. He's tough, and aggressive ...'.

I noticed he walked with a limp. Media files from a couple of decades ago explained that when he was aged twenty-one — before the America's Cup Challenge — he was totally paralysed with spinal injuries after a car crash. The prognosis was that he would not regain use of his arms or legs. But he realised he had some feeling in just one foot, and after six months he was walking and using both arms, to the genuine surprise of surgeons.

Noel Robins' *curriculum vitae* for the *Duyfken* Foundation listed his impressive sporting career: two-ton World Sailing Champion; former National Soling Champion; skipper *Australia*, 1977 America's Cup Challenge; skipper of *Hitchhiker* in two Admiral's Cup competitions; winner of one world, four national and thirteen State yachting titles.

He had been a schoolteacher from 1957 to 1969, and was now a board member of the Western Australia ParaQuad Association and a commissioner of the Western Australia Water and Rivers Commission. In the past he was managing director of Royal Perth Yacht Club's America's Cup 1987 marketing company ADAC Ltd (Australia's Defence of the America's Cup); executive director of that club's 1987 committee; and a property developer, building hotels, home units and nursing homes.

By July 1998 the Foundation had raised about $3 million of its construction target of $3.7 million and the ship appeared to be on schedule for the launching in January 1999. After that would come the final fit-out and rigging, then sailing trials.

In 1996 Adriaan de Jong produced a concept drawing of the proposed old Dutch-style shipyard for potential sponsors of that project. Later in 1996 the Western Australian Lotteries Commission put up $250,000 to sponsor the building, to be named the 'Lotteries *Duyfken* Village'. The deal was designed to give the commission significant visibility in what was now becoming recognised as Australia's biggest and most colourful historical project.

A large number of sponsors were invited to help get the shipyard built. The engineers Entact Clough undertook the big task of bringing the trades together as a group and into the project, and the result was a $140,000 saving in the $310,000 shipyard village project. The group swiftly transformed the rectangle of land outside the Maritime Museum into what would function where necessary as a modern shipyard with concrete footings, pavers and power outlets for the appropriate electrical equipment. In some cases suppliers donated their services while others gave a best-price deal. In total it reflected a great level of community support for the replica project and the enthusiasm and generosity of the people involved.

There were three key elements to the Lotteries *Duyfken* Village. The components of the building were bolted to each other so that the essential building could be removed when the ship was ready for the next procedure of stepping the masts and rigging them. A raised platform was designed near the stern end, for viewing inside the hull when the planks were going on. The great design challenge was to project the feeling of the seventeenth century despite the essential steel construction of the shed.

Left: Project CEO Noel Robins (left) and Foundation board member Charlie Welker.
Right: Generous donor Julius Tahija.

From nearby Mews Road, the M. G. Kailis Group made available a range of woodworking machinery for the shipyard workshops. The shipyard was completed on schedule, in time for the ceremonial laying of the keel in mid-January 1997.

THE NEXT development was to lift the project into a higher gear and link the three nations sharing the seventeenth-century history, when Julius Tahija, an independent Indonesian businessman and friend of Michael Kailis, offered to pay for all the timber for the hull and rigging.

Mr Tahija was born in Surabaya, Java in 1916, and as a boy he spoke Malay with his friends, Javanese with his family and Dutch at school. His parents were not wealthy. At a trades school he learnt a complex code of business behaviour from some Chinese boys with whom he was learning English, and this included the advice: 'Negotiate using facts and logic', and 'to make a deal, all you need is a handshake'.

His father gave him ten cents a day for his tram fare to school. Julius decided to walk part of the way and save some of the fare to use in his own modest business enterprise, selling rice door to door from an ox cart. Later in life this was to lead to an impressive career as a corporate executive working with American, European and Asian transnationals in diverse fields.

In 1937 Tahija joined the Netherlands East Indies army. Soon after the Japanese opened the Pacific War on 7 December 1941 the East Indies came under attack, and Sergeant Julius Tahija was assigned to transport a group of Japanese civilian internees by boat to Fremantle. He then sailed back for Java, but found that the Dutch force there had capitulated on 9 March 1942, so he sailed straight back south to Exmouth to continue the fight.

Early in 1942, as a lieutenant, he sailed about 480 kilometres north from Darwin as commander of a boat with a small force to resist an expected Japanese army landing at the Tanimbar Islands.

After destroying a petrol dump there and taking the local villagers and Dutch to a designated refuge in the jungle, they fought the Japanese invaders and then escaped to set out for Australia in a small boat. It soon capsized, so they swam ashore and departed in another boat. There was no compass aboard, but Tahija with some experience of navigation managed to hold an almost southerly course of about 160 degrees to aim for Darwin. They ran aground on a coral reef, but the rising tide lifted them after several hours and they made it to Bathurst Island, about eighty kilometres short of Darwin.

Tokyo Radio boasted in English that the Japanese capture of the Tanimbar Islands was the last stepping stone for the invasion of Australia. In distant England, Queen Wilhelmina and her Dutch Government-in-exile awarded Lieutenant Tahija the Militaire Willems-Orde, equivalent to the British Victoria Cross.

He joined the Australian top-secret Z Force unit and carried out a number of missions with them. When the war ended, Tahija returned to Batavia/Jakarta, where he was appointed as aide-de-camp to a Dutch general. But the escalating Indonesian war of independence led him to resign in 1946 and go into politics. The same year, he returned to Australia and married Jean, a Melbourne girl he had met during the war.

Several years later Tahija left his government role and went back into the business world. In the 1960s he considered a partnership with a large foreign logging company, but chose instead to become active in support of the World Wildlife Fund. He was popular at home, being politically independent.

On 4 October 1996 he was in Fremantle Harbour again. In the Great Cabin of the *Endeavour* Replica, he accepted the position of Patron, for the Republic of Indonesia, of the *Duyfken* 1606 Replica Foundation. He shook the hand of his friend Kailis — that was all he needed — and the Foundation was thus assured of all the timber required for the Little Dove.

LATE IN 1996 the *Duyfken* project's research and design coordinator Nick Burningham went to Latvia to meet the Unimark Timber Co. representatives there and select the appropriate oak timber for the first load.

By this time a *Duyfken* construction team of four had joined the project, with skills brought from the highly successful *Endeavour* replica project.

The leader, master shipwright Bill Leonard, was a genial Scot who had served his trade time in the British shipbuilding industry. Justly credited with the impressive success of the *Endeavour* construction at Fremantle, he was now recognised as one of the world's leading historic replica builders. Mr Leonard had spent the previous two years restoring small boats for the Western Australia Maritime Museum.

Ship's carpenter Nick Truelove had done all the interior work on the *Endeavour*'s great cabin, the gentlemen's quarters and the cooks' cabin. Josh McLernon had started work on *Endeavour* when he was aged fifteen. He spent six-and-a-half years on the construction, became a topman in charge of watches and ship's maintenance and rigging. Boat builder Don Cockrell had spent many years with the Western Australia Department of Marine and Harbours.

The arrival of the first load of twenty-five tonnes of oak timber at the shipyard on 27 November 1996 was a moment to be remembered. Now we had something to show for our efforts. The first lift of planks from a semi-trailer was a short one, but the next one brought all watching eyes and camera lenses up (including mine) as it rose silhouetted against the bright blue sky and a layer of white cloud. When it landed, Bill Leonard wasted no time in opening up the plastic wrapping to study the timber.

When the unloading was completed, Michael Kailis was proud to take up an adze and make the first formal cut in a piece of this first shipment, symbolising the beginning of construction of Australia's first European ship. The oak proved surprisingly tough, and he swung the traditional shipwright tool again and again to produce a few chips for posterity. 'A small cut, but a giant step for the replica project', I said solemnly as I photographed the chairman in action. 'I expect this day will be remembered by Friends of the *Duyfken* and their children and grandchildren when they watch our ship under full sail off the coast', he said.

Graeme Henderson, director of the Maritime Museum and a member of the Foundation board was watching. Jeremy Green, head of the Museum's maritime archaeology division, came down from his office to see the significant activity relating to an earlier Dutch ship than the proud *Batavia* whose skillfully excavated stern oak timbers reigned majestically within the building.

Robins, chief executive officer of the *Duyfken* project, voiced his excitement: 'Now we can see the vision taking shape...'.

Blessing the keel: Father Vincent Conroy of Fremantle Missions to Seamen conducts the traditional ceremony — pouring holy water from a (copied) historical Cardinal Bellarmine jug.

IN JANUARY 1997 came the opportunity to show off the *Duyfken* project to the world media. The Netherlands' Prince Willem-Alexander, traditional Prince of Orange, had accepted an invitation from the organisers of the de Vlamingh tricentennial celebrations to visit Western Australia to celebrate the event. He also agreed to formally lay the keel of the Little Dove. As it transpired, the combination proved a sparkling success for the *Duyfken* Replica Project, while it enhanced the royal visit and the de Vlamingh celebration tricentennial program rather than steal the glory.

Prince Willem-Alexander with *Duyfken* Foundation Chairman Dr Michael Kailis at the laying of the keel, January 1997.

The essential parts of the replica keel were set up with urgency to meet the deadline, and on Friday, 10 January an ancient maritime ceremony, blessing of the keel timbers, was conducted at the shipyard. Father Vincent Conroy, chaplain of Fremantle's Stella Maris Seafarers' Centre, performed the procedure intended to ensure that the Little Dove was blessed with good fortune both during her construction and when she was eventually launched.

Next morning Prince Willem-Alexander came to the Maritime Museum and the Lotteries *Duyfken* Village Shipyard, where the replica's keel waited, suspended by a block and tackle. The big shed was crammed with about 200 officially invited guests, media representatives, supporters and others lucky enough to be admitted to the limited area.

The crowd rose and shuffled anxiously for glimpses of the tall young Prince, casually dressed in shirt and tie. After the official welcoming speeches were over the Prince smiled at his respectfully silent audience and delivered a short address. 'This project is very important to Holland and Australia', he said in a clear voice with excellent English pronunciation. 'I hope she sails well...We'll see.' He spoke what mattered in a few words.

He moved confidently back toward the stern and his arm could be seen rising to lower the oak timber section into position on the supporting blocks. This was the moment of birth of the Little Dove, I thought. Prince Willem took up the invitation to throw the first bright orange ribbon high above the mock frame of the ship, and amid deafening cheers the air was filled with a gleaming host of the guests' flying orange streamers signifying goodwill to the Prince and pride in the new Little Dove.

THE SUM required to build, launch and complete the *Duyfken* replica rose slightly in a couple of years from an estimated $3.5 million in 1996 to $3.7 million.

By the end of July 1998, the *Duyfken* Replica Foundation's hard-working executive committee, with the responsibility for raising sufficient funds, had raised nearly $3 million. The pledged list included:

Federal Government: $500,000
Western Australia State Government: $500,000
Dutch Government: $385,000
Western Australia Lotteries Commission: $600,000
M. G. Kailis Group: $350,000
Amcor Fibre Packaging: $100,000
Friends of the *Duyfken*: $50,000

Con Smit, chairman of the *Duyfken* Friends, was justly proud of his group's contribution of $50,000 to the project by September 1998. He expected the Friends organisation to continue after the ship was launched, along with the *Duyfken* Guides, who played an important voluntary part in assisting and explaining to the public at the shipyard.

The informal group which was eventually to become the Friends organisation had been formed in early 1994 as a community support group for the *Duyfken* Replica Project when the project's founding members gathered supporters for committee meetings at the home of Michael and Janine Young in East Fremantle.

After Michael Kailis agreed in 1995 to become chairperson of the project, the informal supporting group was transformed into the Friends organisation to operate as a subcommittee, with chairperson Rod Murfitt. The major objective of the Friends was stated to be to raise funds for the project. In October 1996, after the Friends chairperson resigned, retired engineer and company director Con Smit was elected to the position. The enthusiasm of the Friends was impressive, and there were too many consistent workers to be listed for their efforts in organising functions to raise money for the project, but Mr Smit noted a few committee members, vice-chairman Bert Creemers, Nel Ottenhoff, Don Edwards and Geoff Halstead.

Mr Smit recalled that Prime Minister Mr Wim Kok had visited Australia the previous November and at a welcoming State dinner at Parliament House, Canberra, Australia's Prime Minister John Howard had announced a $500,000 contribution to the *Duyfken* Replica Project. The Dutch Prime Minister had endorsed the international significance of the wonderful project and pledged to match it with a gift of 500,000 guilders (about $A385,000).

Not many people were aware that the Netherlands was fourth, after the US, UK and Japan, in direct investments in Australia. The total investments of the Netherlands in Australia were now $12 billion, of which $8 billion was in mining. There were now seventy major Dutch companies active in Australia. Shell was importantly involved in the North-West Shelf gas project.

In April 1998 the Dutch Ambassador Roelof Smit came to the *Duyfken* shipyard to formally present his government's donation of half a million guilders, and he did so with great style. The eyes of the audience were drawn first to the piece of paper he held up, but we observed that the ambassador's eyes were embracing the beautiful shapely lines of our ship. 'Is it not fantastic?', he said, and paused while the crowd of about 180 supporters applauded the replica. 'We always talk about the small *Duyfken*, but she is fantastic, I think.'

It was time for the ambassador's 500,000 guilders gesture: 'The most important element we have kept until now. Here is an envelope which I hand over to Dr Kailis, chairman of your Foundation. Good luck, safe passage with the *Duyfken*' (he pronounced it 'Dife kin').

Dr Kailis commended the navigational feat of the first Dutch who came to Australia. Then he praised the ambassador's help in dealing with Canberra and the corporate world. 'Believe me: to get two prime ministers to give you some money you need somebody up there', he said. 'Without your help, Mr Ambassador, we would not have had the Federal or Dutch Government support.'

THE *DUYFKEN* FOUNDATION invited the world's foremost authority on sixteenth- and seventeenth-century Dutch sailing ships, Ab Hoving, of the Rijksmuseum in Amsterdam, to examine what was being done on the *Duyfken* replica. The chairman had declared: 'We are extremely keen for him to inspect our work. Our aim has always been to build the best Age-of-Discovery replica, a seagoing ship; not just a model for people to look at'.

Mr Hoving visited Fremantle for a week in May 1998 studying the *Duyfken*, the construction methods and plans. His visit to Western Australia, sponsored by Qantas, was designed to help ensure that our replica was built with maximum possible authenticity. He gave his stamp of approval to the replica. He was most impressed to see the ancient shipbuilding techniques used as a research tool to determine the design of a full-sized replica ship, and the courage of master shipwright Bill Leonard and his team was admirable.

'In the Netherlands we are trying something similar, but more cautiously — building to a set of predetermined plans. The unique thing you have here is a willingness to put faith in the ancient techniques and authentic materials.

'It is even more remarkable that the shipbuilding team, many who worked on the *Endeavour* project, have shifted from an English technique to the very different Dutch style of construction of the sixteenth century. In those days the construction method had a large influence on design, something that hasn't been tried in the Netherlands yet.'

The *Duyfken* Foundation's research was also breaking new ground in the area of sailing ability, he said. 'Very little is known about ships of this period, but the studies of Nick Burningham and his research team have demonstrated that *Duyfken* must have been a very creditable sailing vessel. You are the first people I know who have tried to recreate the sailing performance of the original vessel in their replica design.'

Ab Hoving gave a public lecture at the Western Australia Maritime Museum's *Batavia* Gallery and also addressed an enthusiastic group in the *Duyfken* Lotteries Shipyard who were anxious to hear his opinions about the replica of Australia's first European ship of 1606.

Of the three major classical ship reconstructions Holland had produced over the last fifteen years — *Prins Willem*, *Amsterdam* and *Batavia* — two were built on the basis of models and one from not much more than suppositions worked out in a twentieth-century way, he said.

English master shipwrights of the sixteenth century worked in Admiralty shipyards from long lists of measurements, sizes, and dockyard models. But nobody checked the Dutch shipbuilders. They were masters in their field, and the profession had a lot of secrets, some of which would die with them.

Mr Hoving found that the design and building had been one process. The ship gradually emerged in the shipyard under the hands of the person who designed it. The time was surprisingly short: a merchantman forty metres long took four to five months with only twenty to twenty-two skilled carpenters. A capital man-of-war took half a year with about twice as many carpenters.

In the Maritime Museum, Mr Hoving admired the scale model of *Duyfken* built by Dr de Heer, and both commended the lines and progress of the replica.

Later in May the Australian Minister for Communications, Information and the Arts, Senator Richard Alston, visited the shipyard in connection with the Federal Government's $500,000 grant and announced that Australia Post would celebrate the replica by issuing an exclusive commemorative envelope to help all Australians understand the significance of the *Duyfken*'s 1606 voyage.

MEANWHILE, the Foundation's tireless and long-sighted board member Sir James Cruthers kept in mind that raising $3.7m was not the end of the story. At least another $2m would be required to finance the planned voyages, including a voyage from Indonesia to Australia in the year 2000 and later the 1606 re-enactment voyage from the Netherlands to Queensland's Cape York Peninsula.

Sir James came up in early 1998 with a colourful initiative: the 1606 Club, 'Australia's Most Exclusive Club'. Membership would be by invitation only, for a one-off tax-deductible donation of $1606. Benefits would include the following:

- An impressive *Duyfken* 1606 Club membership certificate.
- A quarterly newsletter, the *Duyfken Report*, on the vessel's progress.
- A special pass to the Lotteries *Duyfken* Village and invitations to 1606 Club functions.
- VIP viewing at the launching of *Duyfken* in January 1999.
- A personal tour of the replica after the launching.
- Permanent acknowledgment on an honour board.

He did not need to explain that if the 1606 Club attracted just fifty members in a year or so, that would yield $80,300 for the replica project. It sounded a bit too good to believe, but he set out to make it work, with brilliant results. By August the 1606 Club had sixty-five members ($104,390), and things were looking up, with about twenty more expected in the next few weeks. Chairman Graham Reynolds was optimistic that membership would eventually rise to more than 150, representing $240,900.

Mr Reynolds frankly described the fundraising of the club itself as 'levering' more members to join, or 'collective arm-twisting' by the committee of Sir James Cruthers and Alan Blanckensee, adviser to the Western Australia board of Challenge Bank, and himself. The three met weekly at the Western Australia Club in St George's Terrace, Perth and deliberated on methods of recruiting membership. One part of the strategy was to persuade each member to name three potential new members. It was obvious that Graham Reynolds was enjoying the hard selling for the Little Dove.

As Sir James described it, 'You have to get people into a headlock'. An increasing number of Western Australian yachtsmen were surrendering to the jovial arm-twisting, and he visualised a great increase in $1606 membership when the *Duyfken* replica sailed on its historical re-enactment voyage from Banda to Cape York Peninsula and to the eastern Australian ports.

IN SEPTEMBER 1998 the Foundation set up a committee to begin planning for a re-enactment of the *Duyfken*'s 1606 Australian discovery voyage, and national historic celebrations at Pennefather River and Weipa.

The committee, with chairman Charlie Welker, considered a schedule suggested by Nick Burningham for the replica to depart from Fremantle, probably on New Year's Day 2000, on a 4000-kilometre voyage to Amboyna, taking about four weeks. From there she would sail about 220 kilometres to Banda, where *Duyfken*'s voyage of discovery actually began. She should sail the 520 kilometres to the Kei and Aru Islands in four days, and then begin the 1100-kilometre course to Cape York Peninsula, arriving at Weipa about the end of March 2000.

After celebrations, she would be towed the 2800 kilometres around the top of Cape York and down to Brisbane, arriving there in April. Then the months of May, June and July would be spent sailing up the Queensland coast, stopping at ports such as Bundaberg, Gladstone, Rockhampton, Mackay, Townsville, Cairns, Port Douglas and Cooktown.

From there she would sail 1600 kilometres to Darwin in a couple of weeks and then down the West Australian coast, stopping at Broome, Port Hedland, Dampier, Exmouth and Geraldton, getting south of the tropics before the cyclone season. She should be back in Fremantle for the Christmas holidays.

From the Maritime Museum perspective, the replica project had been an outstanding success, Museum director Graeme Henderson declared in May 1998, when the construction had passed the midway point. He summed it up:

- It has proven that partnership between the Maritime Museum and community foundations can succeed to the benefit of both parties.
- It has attracted media attention to the Maritime Museum and Museum products.
- It has produced revenue for both organisations,
- It is likely that the State Government's observation of this successful partnership encouraged it to proceed with the more ambitious Waterfront Precinct project.
- It has been a valuable trial run for the Maritime Museum's education staff interested in providing a service encompassing the new Waterfront Precinct.

The project, most significantly, was highly authentic, Graeme Henderson said. From his experience in the heritage projects in the Maritime Museum he was aware that credibility was all important — the visitor was expecting to be moved by the experience, and would not be moved unless a product was absolutely convincing. 'The *Endeavour* people recognised this, and I am proud to say that the *Duyfken* people have carried authenticity to a yet higher level.'

The project was responsibly and economically managed. 'We all worried when Bond Corporation faltered and the *Endeavour* project stalled for a year, and the story of it costing $15 million before its keel was wet made it an expensive operation. In contrast the *Duyfken* project is seen as consistently and economically managed.

'Noel Robins as CEO has played a major role in all of the above. I have been particularly impressed by his ability to bring a diverse group into harmony.

'The culture of the heritage shipwrights and the maritime heritage enthusiasts in Fremantle has in past years been characterised by incessant bickering. Noel has his staff working in harmony, and increasingly the other heritage shipwrights and maritime heritage enthusiasts are finding ways of associating with the *Duyfken* Replica Project.

'I look forward with great enthusiasm to the more complex coming days of this most significant project', he concluded.

VII The Shipbuilders

A COMPLETE description of the 1997–99 *Duyfken* reconstruction was beyond the scope and timing of this book. Among other factors, in July 1998 the project was about three-quarters finished, with decisions continuing to be made before the ship could be ready for launching in January 1999.

The scale model built by Dr de Heer had inspired the creators of the *Duyfken* Replica Project, but the initial plans from the beginning of the project underwent a number of changes, including the shape of the bow and the number of decks. She was described as a jacht, or yacht.[51]

The design evolved through research and dialogue between a number of people. In Australia these included master shipwright Bill Leonard, nautical archaeologist Nick Burningham, researchers Adriaan de Jong and Marit van Huystee, model-maker Cornelis de Heer and maritime archaeology division head Jeremy Green, while in the Netherlands there were Thijs Maarleveld, Ab Hoving and Robert Parthesius.

Some of the information was to be derived from historical records of the events in which our historic ship was involved and also from the study of drawings of the Little Dove and other ships, by artists of their time — 'morphometric analysis of the iconography' as Nick Burningham would shudderingly describe it. The process would necessarily be experimental because there were some questions about the design and building methods that might only be answered as the work progressed.

While some traditional shipwrights' tools and methods such as the adze and fire-bending were used in the construction to help maintain the *Duyfken*'s authenticity, the design and planning process took advantage of sophisticated computer analyses and other scientific aids in the aim to produce a worthy sailing ship.

To adequately describe the fascinating but complicated undertaking, it seemed best to follow the story from the perspectives of several of the individuals most significantly involved in the design and construction.

Bill Leonard:

Bill Leonard built the replica of the eighteenth-century *Endeavour*, which was internationally acclaimed, especially at the English port of Whitby where the original had been built. This replica was constructed using original eighteenth-century plans. For the *Duyfken*, Leonard's skill, experience and especially his expert eye were to be tested again.

Born on the Firth of Clyde, Scotland, he served his first three years of apprenticeship with a wooden boat building company called Fairley Yacht Slip, on the coast south-west of Glasgow. This was a large yacht building firm, famed for the America's Cup yachts it built, owned by William Fife of Fairley. The fourth year Bill spent at a small shipyard, John Swinbourne's in north-west England, building clinker dinghies, and the last year he worked with the Swan Hunters shipyard in England, building steel ships.

His apprenticeship allowed him a wide range of construction experience, from quality yacht building through to large wooden fishing boats, down to little clinker dinghies and then in his final year building large steel ships. In terms of the shipwright's trade, it closed the circle.

He came to Western Australia in 1986: 'With the America's Cup coming up I knew I could find plenty of employment here. I got in touch with Fremantle Shipwrighting Co. Western Australia Pty Ltd, who advertised in the Yellow Pages as builders of fine wooden ships. I thought that was just my role, and they gave me my first start in Australia.

'I was introduced to Steve Ward and I got involved with building twelve-metre yachts ... Steve Ward had the initial contract to build the *Endeavour*. After the twelve-metres I was about to pick up my bat and ball and go somewhere else, and Steve Ward said "Don't do that: you're a builder of wooden ships". So as soon as the *Endeavour* came on stream I was involved in doing all the lofting for it. It's the first shipwrighting job in building a ship.'

After the *Endeavour* replica was completed Leonard spent several years working with the Western Australia Maritime Museum, which was respected internationally for its studies in maritime archaeology and its excavations and conservation work on Dutch and colonial shipwrecks. When he eventually accepted the two-year contract to build the *Duyfken*, Leonard was aware that some unusual constructional problems would have to be solved.

'We probably know more about the construction techniques of the Roman shipwrights than of the nuts and bolts of the everyday ship construction that took place in the sixteenth-century shipyard', he said in February 1997, in an address to supporters of the project. 'Hopefully the *Duyfken* will answer the questions that the sometimes abstract world of maritime archaeology asks.'

What were the available clues? 'A few notes, some drawings, books of formulae, a few interesting contracts and, of course, those magnificent fragments of the early Dutch shipwrights' works — the most famous in the Southern Hemisphere, the *Batavia*.'

During Leonard's periods of working with the Maritime Museum, he had been intrigued to observe people entering the *Batavia* Gallery with its towering exhibit of the stern timbers of that Dutch 1629 shipwreck.

'They would wander in to be confronted by a wall of wood with interesting little openings near the top, and then they'd walk around to the foreside or look down from the gallery level above. What did they see? A confusion of random shapes, shambles of wood piled upon each other.

'What they viewed was the art of the ancient shipwright's trade. To modern shipwrights like myself, these magnificent relics of ships built by master shipwrights give the answer to HOW.'

Some notable master shipwrights in sixteenth and seventeenth centuries were people of substance, highly educated and organised; for example, the famous sixteenth-century English shipwright Phineas Pett. Born in 1570, he started boarding school at Rochester and later in his teens at Greenwich before completing his four-year degree course in 1590 at Emmanuel College, Cambridge. Then he was fortunate to acquire a shipwright apprenticeship at Debtford, eventually in 1595 falling in with the famous Matthew Baker, whose ancient ships' drafts are available for all to see in English archives.

So how did the master shipwrights start to build a ship in the sixteenth century? 'By first of all having in their hand a contract.'

He set the historical scene: The great port of Amsterdam, crowded with hundreds of sailing ships, overlooked by the 'Weepers Tower', where widows might wave goodbye or grieve for their lost sailor husbands; the first Indies fleet of Cornelis de Houtman had sailed from near this tower in 1595 ... the busy city of canals and elegant buildings of many gables ... a house with insignia on the entrance denoting the profession of its occupant, the Master Shipwright.

A knock on the door, and there before the shipwright stands a wealthy merchant. He treats this skilled builder of ships with respect and candour.

The merchant begins: 'Master Shipwright, I need a ship of great burden. Of course it will be grander than that old hoeker you see over there just returned from the Indies. This ship will reflect my position in the City while I make my fame and fortune in the East Indies'.

The master shipwright understands. Before long he reaches for a piece of parchment and then turns to his pen and inkwell to quote his price for the merchant.

On recovering his breath, the merchant exclaims in his best double-Dutch: 'Sir, you charge like a wounded steer'.

The shipwright responds with appropriate indignation but eventually relaxes into a frank Dutch entrepreneurial appraisal of the situation: 'I care not for fame, Sinjeur, and I must make my fortune right here'.

Centuries had come and gone but some things never changed, Bill Leonard commented in his Scottish tones, slipping, perhaps appropriately from now on, between past and present tenses, and the next question from his Dutch merchant was the big one: 'What do I get for my florins?'.

In early 1997 several sixteenth-century Dutch shipbuilding contracts were available for study, and more would turn up later. Contracts are a really important clue for the building process of early ships because they often describe the construction in a sequence of events — the scantlings of materials, how they were joined, fastenings to be used. It helped a bit to be able to read a little sixteenth-century Dutch.

Leonard was glad to have some information available nearby from the Dutch ship *Batavia*, said to have taken only nine months from the start to her time of launching. 'That shows very real organisational skills, not a confused process of construction, not a haphazard shambles, but an amazing degree of pre-planning, a complete understanding of the shipbuilding process, a well organised and controlled labour force. They knew exactly what they were doing — and knew that mistakes cost money.'

The sixteenth-century builder, with his contract in hand, would have met with other important people, suppliers of wood, sparmaker, blockmaker, sailmakers, riggers, blacksmiths, suppliers of tallow, tar and oils. His thoughts would settle on the keel, and probably the number of keel blocks required, the condition of the stocks (the slipway foundations) that the keel blocks would be secured to.

Surprisingly at this stage, a launching procedure is clarified in the Dutch master shipwright's mind. His previous experience and intuition are coming to the fore: no point in building the ship if he can't get it into the water.

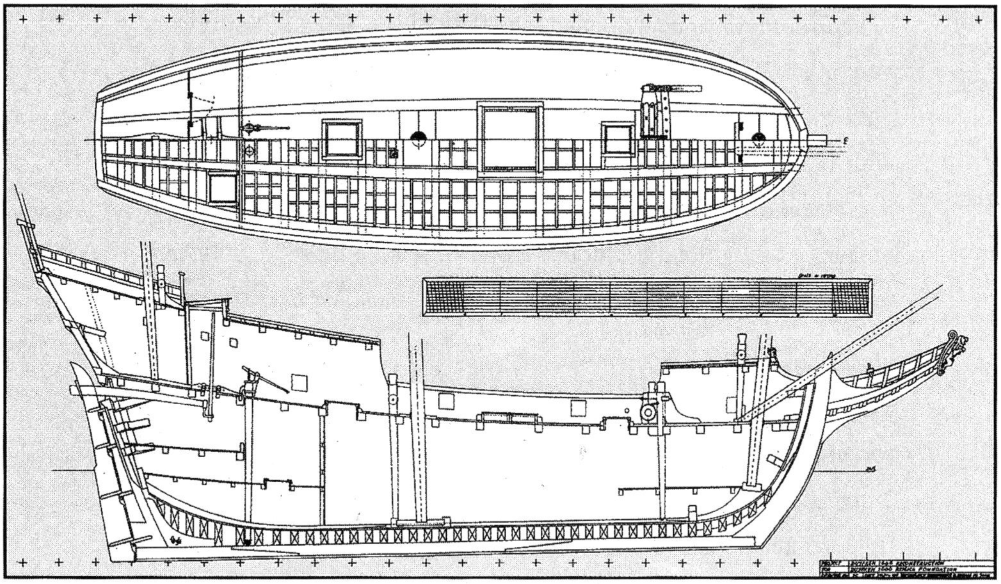

Duyfken replica plans: side elevation and deck plan, by Adriaan de Jong.

Soon the sawyers in the pits start on timber for the keel, the dimensions of which were given by the shipwright from his book of proportions and formulae. Perhaps these dimensions were passed down from his father, but his personal experience would guide him. Perfect straight cuts reflect the sawyer's skill. The shipwright, joining these pieces together with a tapering scarf joint, then proceeds to cut out the rabbet, or rebate, the recess in the keel that takes the lowest plank, the garboard strake.

The bevels and angles are known to the shipwright, the final trimming determined by the twist of the plank itself as it reaches the stem and sternpost. The shipwright could well be concerned about the supply of good compass (curved) timber. He certainly cannot build the ship without it, and having to select it himself, he will find it a time-consuming process. He needs it by the cartload.

The first major piece of compass timber required is the stem, and its shape is determined by a couple of sweeps of a line, or in many cases the curvature of the timber itself. On larger ships the stem would have been built of two pieces. Down at the other end of the keel, the shipwrights have started the sternpost framing, consisting of the sternpost itself and compass timber in the form of fashion pieces, wing transom, lighter pieces of wood in the form of stanchions. The long narrow top timbers extend this prefabrication right up to the stanchions on the upper deck.

This large structure was not only a major structural element of the ship but also defined the shape of the stern. Like the stem, it was secured to the keel with the help of large knees fastened in place with bolts. When the stem was set up on the keel, the major centre line structure of the ship was completed.

During this period, ships constructed in the Netherlands were built shell-first and planked out to the turn of the bilge (the nearly horizontal part of a ship's bottom) before any framing timber was inserted.

The garboard strake at the bow tended to be sharply twisted and bent, and would have been formed, with the help of an adze, from naturally curved timber and then heated over a fire. To achieve its final curve, the shipwrights used an iron device, a sort of adjustable fork, with which they twisted the plank. One end of a plank would be secured to the bending frame and, while a fire was burning beneath it, the iron device was used to slowly twist the planks into the required shape.

The garboard would then be nailed into the rabbet on the keel and the stem and sternpost, while cleats on the keel ensured correct angles between the keel and the strake along its length.

The fire-bending of planks would continue, the next strake being fastened with temporary cleats to the garboard. Props (shipwright called them 'shores') were also employed to help create the proper shape. As the planking extended out from the keel, clamps and chains were used to pull the planking together, and this method of construction, if applied by skilful shipwrights, could produce wholly satisfactory results.

Symmetry would be checked constantly by the use of a long plank and a string level extending over both sides of the ship.

When the planking had reached out to the turn of the bilge — the start of a turn toward vertical in a cross-section of the ship — the initial floor timbers would be fitted, running across the ship, and a few first futtocks (from 'foot hooks' — middle timbers between the floor and the top timbers) would be installed.

After all floor timbers were fitted, the keelson — a heavy beam fastening the ship's floor timbers to the keel — would be put in place. When the ship had been planked up through the turn of the bilge, second futtocks would be installed which would extend upward to the full height of the ship's sheer (the upward slope of the ship's lines toward bow and stern). They would then be interconnected fore and aft by a rib band that extended from the fashion piece top-timbers in the stern, to the stem itself.

These second futtocks (and later the top timbers) were set up ahead of the planking, to define the shape of the upper hull. By now it would be something that looked like a ship. After the remaining framing up, including apron hawse-pieces (framing around the bow), counterframing, fitting of the upper part of the stern transoms, and fairing, the planking would continue apace to the sheer.

The fastenings then used were trunnels (from 'tree nails', lengths of dowelling) with wedges driven in each end, to fasten the external planking through to the ceiling planks (internal planking of the hull). The remaining transom knees, mast steps and so forth could now be installed.

Master shipwright Bill Leonard.

As the internal planking extended up, it met the clamps, which were in effect heavier planks that supported the deck beams. These clamps, or shelves, went in before the rest of the internal planking. The construction of the deck would involve laying the lightly curved deck beams and further structural members such as carlings, binding strakes and ledges. The deck was laid, starting with the heavier waterways. The completed structure was caulked with oakum, the seam being finally paid with pitch.

In the meantime, other important components were being made for the ship — the rudder and whipstaff (not a wheel), windlass, beakhead, elmtree pumps, hatches and gratings.

Back in Leonard's continuing sixteenth-century scenario, the internal fit-out started then to loom in importance for the Dutch master shipwright. The yard blacksmith and his team of helpers would have been busy during the construction process with the supply of spikes, nails, ringbolts, hinges, hoops for carriages, rings for spars, forelock bolts, perhaps even the ship's anchors. The sparmakers would have converted trees into spars. The sailmakers might complete the sails after the launching. The riggers, having completed as much preparation as possible, were waiting in anticipation of a successful launch. After that, they could step the masts and rig the ship for her working life.

Before the launch, the hull would likely have been tarred and the underwater parts of the hull would probably be dressed with a mixture of pitch, tar and resin or white lead to prevent and lessen fouling. The stocks (on which the ship rested) would have been greased with tallow or whale oil, and the ship could finally be launched.

In February 1998 Leonard contemplated the *Duyfken* reconstruction, past the halfway point and already revealing quite pleasing lines, as all viewers seemed to agree. She was scheduled to be launched in January 1999. The final fit-out would be completed and the ship's master would take over from the master shipwright.

He had tried to stay a couple of weeks ahead of everybody in his work plan. 'There's a pressing need for some large structural members to be put into the ship. If they come in time, the deck will indeed go on in about three months time. But it all depends. I've got to find all the large hanging knees for the deck beams. I need forty of those.'

Leonard explained the next stage in the work: 'Once the deck beams are installed, then we'll prepare trimmings: carlings and ledges. The carlings run fore and aft, and the ledges sit on top of them. Before the ledges go in, we will be installing the waterways which run around the ship, and then the tying strakes, to support the decking right through the middle of the ship. These tying strakes are extra thick planks that go under the deck.

'When the tying strakes are installed, we can then fit the ledges and start to lay the deck. But before we lay the deck, we've got to ensure that all the fittings which penetrate the deck are already in place. Elmtree pumps, windlass bitts, the carrick bitts for the windlass, the knights — these pass right through the deck. All these heavy structures have to be in place. The ceiling (internal) planks of the ship have to be completed and trunnelled. Once the deck gets on, it's dark in there, and airless. So you always build a ship from the bottom up.

'One of the things we have to do here is to find a flow or pathway through these various components.

Fire-bending one of the wales.

'From there, when we've installed the waterways, put in the tying strakes, laid the deck, we've got to fit out the waist of the ship, the internal lining above the deck. On top of the waterways there is to be what we call spirketting, which is an extra thick planking inside between the top of the waterways and lower sills of ports.

Lowering the keelson into position.

'I'm going to install the waterways and the spirketting in such a way that it ties the sides of the ship together. Once the spirketting is in place, we'll be almost up to the gunports. So with the planking fitted in between the gunports, the ship will be planked right up to the top of the quarterdeck. We can be thinking about installing those deck beams as well.'

The stern castle and forecastle were yet to be erected.

'Now to stepping the masts: the forward mast step is actually part of the keelson, up in the bow of the ship. The foremast is stepped right at the stem on a strake which goes across the ship to form an internal floor. The mainmast step sits on top of the keelson, supported by riders that run out over the ceiling.

'We have the wood for the beakhead. It's a double curve, a nice piece of wood. There's another six months to go before we can install the beakhead because it would get in the way of the crane and we couldn't lift anything off. Jenny Scrayen is doing the carvings for it. We may have problems squeezing all the carvings in there. There's plenty of carving for her to do, and it's time-consuming work. About that time, I expect, the top will be coming off the shed.

'There's a lot of fitting out to do. Once the deck is on, the ship's a long way from being finished, but after we've caulked the deck and the hull, we can launch her. The aim is to have everything visually completed by the time of the launching.

'My two-year contract with the *Duyfken* Foundation comes off in January. So the ship will be jolly near completed by then. But she must go through a process of commissioning, and everything has got to work; any problems have to be fixed. You can't find all the problems until she's afloat and the spars are all on and she's sailing.

'Once the ship has been commissioned, she's then accepted by the master or whoever's in charge of her. Then it's he who has to make decisions about how to sail her. Once the ship's master comes aboard, lines of command follow.'

'The ship is at a very exciting working stage of construction now', Leonard said. 'I've got a fine team around me now, loyal and very hard-working. It's a pleasure to work with them. We have five tradesmen and four apprentices and a lot of trade assistants and very good craftsmen involved. There must be twelve people working here on the ship at any one time now. They keep me running from point to point, so much that I go home at night feeling knackered. But it's a really good team of people, and it's a happy working squad.'

A couple of lines of poetry could best express Bill Leonard's feelings:

> Ships are the nearest things to dreams
> that hands have ever made.[52]

Igor Bjorksten:

Bill Leonard told me about a man who had been following him around for weeks as he worked on the *Duyfken*. 'Each time I cut a piece of oak he smiles and carries away the off-cut. He's been helping me install the top timbers, and by now he's amassed a pile of wooden blocks.'

It was Igor Bjorksten, *Duyfken*'s master rigger. Later he was to retire from other construction activities and just concentrate on his primary role of rigging the ship. The 1606 *Duyfken* was a small ship with a relatively simple rig, Leonard said, but even so, it was a job for a competent professional to set it up today. 'The master rigger will use his pile of oak off-cuts to make rigging blocks (pulleys) for the ship, fashion the ropes, get everything ready for when she is launched. The masts have already been worked at. Lumps of oak suitable for blocks and trunnels have been sent south of Fremantle to an air drying kiln at Bunnings Timber Co. A short spell of wet weather made them too wet for use as trunnels or blocks.'

Mr Bjorksten, from Melbourne, had rigged ships in Australia and around the world before working as principal rigger for the *Endeavour* replica at Fremantle. When Bill Leonard wanted him for the *Duyfken* project he found that he had gone away to work on other ships, but he traced Igor Bjorksten from Victoria to Scandinavia where he was rigging and sailing square-rigged ships. There were a few square-riggers sailing there, but Mr Leonard persuaded him to come to Fremantle to rig *Duyfken*.

Three generations back, the Bjorksten family came from Sweden, and Igor was born in Melbourne. His father was a sheep farmer. He had recently finished a year in Europe working in his craft, he told me in a discussion of his role.

All of the spars for the *Duyfken* were roughed out in preparation for their fitting, he said. They awaited the finishing touches — the lower mast yet had no cheeks or cross-trees, trestle-trees, mast caps, tenon on the bottom … those sorts of things.

There was a lot to do in the allocated time. Blocks had to be made up from his pile of wooden blocks, the off-cuts from the shipwrights' work — there were about eighty of these — and he had made up about fifty dead-eyes, used for tensioning the shrouds and stays, to be placed just below the ratlines (small lines like ladder rungs).

'So before any of the rigging can take place, all that associated hardware has to be produced. Things like the chain plates have to be resolved, how they're going to be made up.'

The rope was on order from Holland. 'Most of Scandinavia grows hemp. The funny thing about hemp is that, with the strain you make rope from, you'd have to smoke about a kilo of it to get any effect. There's a lot of paranoia attached to it.'

Master rigger Igor Bjorksten and some of his many blocks.

The sails would be flax canvas. As for the yards, after being delivered as tree trunks they were made into square lengths of timber with a taper, the corners were knocked off, down to eight sides, then sixteen, thirty-two and sixty-four sides, just about round.

He indicated on a ship model the position of the 'rahout' (yard rail) on the top timbers. 'You would literally lower the yard down to the rail. With the topsails, most of the furling is done in against the mast rather than out on the yard. Apparently in this particular ship, the topsail yards are sent entirely down to deck. So there would be no furling aloft of any sails. The course yards would come to deck and the topsail yards would be sent entirely down.'

As ships got bigger it was obviously a greater exercise both in hoisting a yard and lowering it, furling it and so on, he explained. But not in this period. The angle of the yard of the mizzen sail was about forty-five degrees. The spritsail was a square shape, but that yard would be sent out from the beakhead.

It was a relatively simple rig, Bjorksten said. But fifty years later they got into really fancy solutions to spread the load out. Another fifty years on, they completely simplified it and changed their methods. The next really big technological change, after 1700, was the introduction of wire about 1860.

I was curious about where the lookout man would be stationed. 'Probably that point there', he indicated on the foremast about ten metres above the waterline. How long would it take, after *Duyfken* was launched, to complete the rigging? 'I can more or less prefabricate the whole thing. Without putting the masts into the ship I could produce eighty to ninety per cent of the rigging, outside the ship. I would make up all the shrouds but not turn the ends up. Shrouds are ropes forming part of the standing rigging and supporting the mast or topmast. Otherwise all the different components can be made up in the shed. If they wanted to leave the roof on the shed until launch time it wouldn't bother me.'

Research and design coordinator Nick Burningham.

He had calculated that there would be about six months of rope work alone for two people. Much of this would be done before the launch. But he would still be involved in the sea trials to make sure everything worked.

The spritsail and the mizzen would be the two most important sails for any manoeuvring, because they were at each end of the ship. The others were the main driving component.

The problems were worked out 400 years ago. If they didn't make something strong enough then and it blew apart, they just made the next one bigger. All these things were part of a process of evolution. These ships were the absolute height of technology of the day, and they obviously worked. Since their time, there were no major changes to the square-riggers of 1900, apart from the lead of a few stays and individual detailing and the materials used.

The master rigger summed up his craft for the rigging and sails: 'It's my commitment to rig it up and see that it works. I guess it's what transforms a hull into a sailing ship'.

Nick Burningham:

I asked Nick Burningham, the project's research and design coordinator, for his perspective on the designing and construction of the Duyfken *replica. In his words:*

WHEN THE *Duyfken* project's co-founders James Henderson, Michael Young and Kees [Cornelis] de Heer began to widen the group of people involved in the project I was keen to play a role.

At the time I was contracted by the Western Australian Maritime Museum's archaeology department, headed by Jeremy Green, to build a one-tenth scale model of the seventeenth-century ship *Batavia*. Working with Jeremy, I was building the model in a public display gallery using the Dutch plank-first system of construction. Apart from being in an interactive display I was conducting a research project — testing how that system of construction might influence the shape of the hull and some of the theorised advantages and disadvantages of plank-first construction.

The possibility of doing the same kind of research at full-size was tantalising. That kind of testing of theories and ideas about life and activities in the past by recreating those activities in real life is sometimes called 'experimental archaeology'. It is a controversial mode of research, not least where it involves the expense of building replicas or reconstructions of sailing craft.

At that stage the *Duyfken* project was gaining momentum and public support for a range of reasons that had nothing to do with my research interests. The challenge for maritime archaeology was to enhance the project through good research and experimental archaeology but without greatly increasing costs and difficulties. Before the project could be costed and planned in any detail we needed to know the size of the original ship.

My colleague Marit van Huystee was involved in groundbreaking historical research about *Duyfken*. It had long been assumed that the *Duyfken* which made the historic voyage to Australia was the same *Duyfken* that had sailed to the Indies in de Houtman's fleet of 1595. Marit showed that could not be the case and gradually delineated a faint ghost of the real 1606 jacht. Nowhere were the dimensions of that ship recorded.

Estimates of her size had been made, but the more we looked at them the more it seemed that they were based on unsafe assumption and that this was part of a more general problem with replicas of Age-of-Discovery ships. Several replicas of the ships of Columbus and two or three of Sir Francis Drake's *Golden Hind* are among these. They look romantic and some are the result of a good deal of painstaking research, but in general they don't sail properly. Comparison with the logbooks of the original ships shows that the originals were faster and more manoeuvrable.

Nautical archaeology has proposed strict use of the terms 'replica', 'reconstruction' and 'restoration' to designate recreations of ancient watercraft based on different types of evidence and different levels of authenticity. Our *Duyfken* is a 'reconstruction' according to this system but the term 'replica' has broad popular usage and does not create the misapprehension that we are reassembling parts of the original ship.

The basic problem is that there are no authentic plans of any ship from the Age of Discovery because no shipwrights used plans as we understand them in those days. The plans were in the shipwrights' heads. In reconstructing those lost designs we seem to have paid too little attention to performance, and perhaps we have been too willing to use the more plentiful evidence from later times. Reconstructions of sixteenth-century ships from northern Europe have been given the long, narrow, boxy shape of late seventeenth-century ships which is inappropriate and unsafe when combined with the high aftercastle and large billowing sails of the earlier vessels. We wanted to build a replica that would sail as well as the original ship did.

If we were to build the *Duyfken* exactly as the original ship had been built, no designing would be necessary beyond deciding the basic dimensions. A contract would be drawn up to specify the 'scantlings' or sizes of timbers used so that the builder would not be tempted to cut costs by using undersized materials and the shipwrights would then build the ship according to their experience to best suit its intended function. A jacht such as *Duyfken* would be shaped for speed and manoeuvrability and with a gundeck high enough above the waterline for use in the open sea but not high enough to affect stability and sailing performance.

Robert Parthesius, a Dutch maritime historian, provided three examples of contracts for building jachts, but without a sixteenth-century Dutch shipwright to interpret them we needed to commit detailed plans to paper before starting construction. At that stage Ab Hoving, head ship-model restorer at the Rijksmuseum, Amsterdam, and author of the authoritative study of Dutch seventeenth-century shipbuilding, began a lively involvement in the research. His initial advice was none-too-encouraging. 'Everything before 1630 is prehistory', he wrote, and suggested that what we were trying to do was a little like trying to reconstruct the design of a Model T Ford from a torn picture of a stage coach and the hub cap of a Volkswagen Beetle! More particularly he warned that some of the proposals he had seen looked 'too English'.

We had been compiling a catalogue, or database, of reproductions of Dutch marine art from *Duyfken*'s time. This 'iconography' contains a huge amount of important information but it should not be read uncritically. It contains various distortions and omissions, some conscious, some unconscious, yet it could give us the Dutch styling and design if analysed correctly. Adriaan de Jong and I employed a system known as morphometric analysis.

I had used this technique ten years earlier to design a replica of a Macassan prahu as part of Australia's bicentennial celebrations. A similar process is used in the natural sciences by taxonomists to distinguish different species, and it depends on measuring ratios of proportions. Where a large sample of iconography is available statistical techniques can be used to refine the results and sometimes to filter out distortions.

It was this process that showed us not only the distinctive Dutch style of *Duyfken*'s time but also a hull shape quite different from the boxy design of the later seventeenth century, and this was quite a break-through in our quest for a swift and sea-worthy design. Unfortunately the iconography can tell us almost nothing about the underwater body of the ships since it is not shown. Like previous researchers we assumed (with little real evidence) the nearly flat-bottomed shape that seems to have been common a century later.

The next breakthrough was evidence from Thijs Maarleveld, head of the Netherlands Underwater Archaeology Service, which showed that our ship could have a better sailing shape. This was based on the shape of a recently discovered shipwreck from the late sixteenth century, known by its site location as SO1. Once we had the hard evidence of archaeology we saw that there were also contemporary models and some details in shipbuilding contracts indicating a similar shape.

Our plans for the ship evolved gradually. There is always more research that one could do, but it was important that the research never delay the construction. Well before the construction got under way we had to come up with a detailed list of the timber required.

Some aspects of the design were not worked out on paper. We could see from the iconography that a wide range of stem shapes were used, so rather than prescribe a particular shape, we decided that the timber suppliers in Latvia would look for a suitably bent tree and we would let the shape of the tree define the shape of our stem. That was, no doubt, how it was done in *Duyfken*'s time, although with much more wooden shipbuilding going on, there would have been a selection of potential stems to choose from.

Our principle was that no aspect of the design was fixed until it was built. Practical shipwrighting, the properties of the timber and any new evidence could modify the design during the construction. As it turned out, practical shipwrighting and properties of the timber modified the design right from the very first stage. Archaeology and seventeenth-century treatises on shipbuilding agreed about the methods of joining baulks of oak using stepped scarfs sloping upwards toward the bow to make the ship's keel. But in our keel timbers a patch of rot could be cut out if we reversed the direction of one of the scarfs. Fortunately the reversed scarf also made smoother the keel laying ceremony at which the Crown Prince of the Netherlands, Willem-Alexander, officiated.

The research and design work takes time and needs the meticulous approach that my colleague Adriaan de Jong epitomises, but it is in some ways the easy part of the replica project.

In the *Duyfken* project our shipwrights led by Bill Leonard were faced with an unusually daunting challenge. *Duyfken* was to be built in a completely different way from the British tradition of which Bill had become a master. The board of the *Duyfken* Replica Foundation had courageously backed our proposed experimental archaeology and maximum authenticity approach: *Duyfken* was to be built plank-first.

Plank-first construction was a little understood process in the modern, western world. The recovery and study by Jeremy Green of the remains of the Dutch East India Company ship *Batavia*, wrecked on the Western Australian coast in 1629, had contributed to a complete reappraisal of Dutch shipbuilding traditions. During the sixteenth and seventeenth century the Dutch had dominated merchant shipping in Europe. It was well accepted that Dutch ships had been uniquely efficient. It had been largely forgotten that they were not built by the frame-first construction process that many historians believed to have been an innovation that made possible the European Age of Discovery.

As shown by the remains of *Batavia*, Dutch shipwrights had stuck to a more ancient conception of how a ship was assembled. Once the keel, stem and sternpost were set up, the shipwrights began forming the ship's outer shell, plank by plank, with no frames or moulds to predetermine the shape. This allowed them to

Bow: stempost, forward crotches, a few first futtocks in place, some cleats holding planks in place.

work by eye and gave them the flexibility to create whatever hull shape they thought best suited to the trade and conditions the ship was being built for. By contrast, in the frame-first construction processes used elsewhere in Europe, simple geometrical principles were employed to design the shape of the frames, and this restricted the range of shapes that could be built.

If *Duyfken* was to have an authentic hull form she had to be built in that plank-first tradition, and she had to be built from the authentic material which was oak from northern Europe. To build a ship plank by plank with no frames to determine the shape, the shipwrights need to bend each plank before it is fitted, and this is done by heating the plank over an open fire until it becomes plastic, then twisting and bending it to the desired form. This is all done 'by eye' and at the risk of burning the valuable slabs of oak.

For shipwrights trained to modern standards of precision, a return to the craft of plank-first construction was a huge leap in the dark. Not all of Bill Leonard's team were confident that it could work. As it turned out, not all of them could make it work: some had real difficulty conceptualising the shapes required.

The keel was laid in January 1997, then the stem and sternposts were erected. It was May before the planking began. Three months later my diary records a mood of frustration and growing crisis — planking was progressing very slowly and the learning curve seemed to be very flat. Some things were being learned: It was important to get a plank with a lot of bend or twist in place while it was still hot, and it was best to cut the plank approximately to shape before bending because sometimes in cutting the scarf the plank seemed to straighten. The shipwrights were undoubtedly working with more concern for precision and symmetry of the hull than would have been the case 400 years ago.

The first optimistic note in my diary appeared on 10 August 1997:

> Nick Truelove agrees that plank bending is going better now. He ascribes it partly to having someone (Steven Edwards) who is dedicated (in the technical sense) to fire bending. Rather than running backwards and forwards from other jobs when planks are being built Steven stays with them. Result is better bends, less burning of the planks.

There was more to it than that. Steven was more careful in keeping the fires burning evenly with uniformly chopped firewood and he had the patience to let the planks heat until they bent reasonably easily rather than forcing them to bend using levers and pulleys and then trying to set the bend by applying heat. Simpler was definitely better.

Meanwhile huge piles of twisted tree limbs were being slabbed by Rick and Jules Martin to provide the frame timbers that would strengthen the slowly developing plank shell. There were some fifty pieces that had been selected as suitable for making the 'floors' — the frame timbers that stretch right across the bottom of the hull — each has a particular shape ranging from a shallow 'V' amidships to an acute 'Y' in the bow and stern. In the old days a shipwright would probably take a timber from the top of the pile, eye it up and decide approximately where in the ship it might fit, then rough it out to the finished shape and trial fit it. After a bit of fairing off it would be fitted into the plank shell.

Maritime Museum Education Officer George Trotter inspects one of the scores of knees being shaped.

The end result of that haphazard approach can be seen in the shipwreck archaeology: the floors were not evenly spaced and some of them were not quite square to the hull. A more ordered approach was wanted by our shipwrights but it required more searching through piles of heavy timbers and making plywood templates to determine the shapes required at each floor position.

By the time our ship was planked out to the turn of the bilge and some of the floors were fitted, the shipwrights were more confident of the process and could see that it was working, but we were building a shape of ship that they had never seen before so it was necessary to make a set of temporary moulds (templates of the cross-section shape). These were never used to actually shape the planking — they were too flimsy — just to visualise where the planking was going.

In September planking was really going well and the shipwrights were enjoying the creative process. However, it was time to concentrate on getting all the floors fitted and the keelson laid over the top of them. The keelson, which forms a girder with the keel, is the most massive timber in the ship and traditionally needed to be lifted into the hull before the planking had progressed too high.

On 5 September 1997 I noted:

> We have the International Congress of Maritime Museums' conference here these last three days. The delegates are amazed by the speed of progress. 'It's like it's growing organically', one of them said. And it is. Planking is finally going extremely well and, I hope, fitting of the floors is speeding up too.
>
> The beauty of the shape is generally agreed on.

A month later the planking was complete right through the turn of the bilge and some first futtocks were being fitted. At that stage in the Dutch tradition the critical underwater shape of the hull is complete and construction switches to frame-first with the futtocks developing the shape ahead of the planking.

The next great challenge in the construction was shaping the first wale. Wales are double-thickness planks that strengthen and protect the hull at certain points. The first wale runs from bow to stern at the widest point of the hull, near the waterline. The shipwrights had become good at bending sixty-millimetre-thick planking but we were anxious to find out about bending 120-millimetre-thick wales through the tight curve at the bow. Adriaan had shown me a historical document that complains about planks being burnt 'to a shame' in the bending process. There were a couple of failures in bending the wales, but in the end it was done.

Again, the *Duyfken* replica team faced a difficulty that the original shipwrights would have thought most extraordinary. It was near mid-summer when the wales were fitted and some days the temperature in the shade was around forty degrees Celsius, making Steven's job tending the fire very unpleasant.

By early February the whole shape of the hull up to the height of the rail was defined by top timbers or top futtocks, and Gerard Russell began cutting the camber (curve) of the main deck beams. The layout of the decks had been the subject of much discussion.

During the framing the top timbers were fastened to the planking with coach screws where the sixteenth-century shipwrights would have used a few iron spikes. The majority of the fastenings were to be trunnels or treenails. There are thousands of trunnels in the hull. At the end of July trunnelling had been going on for six months, most of the twenty-five-millimetre holes having been drilled and the trunnels driven home by Jim Lucas, a former colonel of the US Marines. It takes a steady hand with the maul (shipwright's sledge hammer) to drive those tight-fitting trunnels.

Ab Hoving's visit to the *Duyfken* in April 1998 provided reassurance that we were doing it right, even some aspects that I was diffident about. The top timbers were being made from grown

Looking forward to the bow, with apron and hawsepieces, deck beams in place, also carlings (part of the lattice supporting the deck), and now the sheer can be seen.

oak limbs of approximately suitable shape, and quite a number of them had to be built up to the required profile with fillers and chocks. Ab Hoving was delighted: 'Just what you see and read about in the old shipyards', he said. More importantly, he was very impressed by the courageous way Bill Leonard and his team were prepared to try the ancient techniques of building and designing a ship. Ab Hoving's comments seemed to strengthen Bill Leonard's commitment to the traditional ways. A week later I noted:

> In a short discussion we eliminated the most forward deck beam, which is too close to the breasthook to do much, and added some foremast partner structure. This is good traditional designing where the master shipwright works it out simply from basic principles and prior experience.

By mid-May 1998 all the main deck beams were in place and it was clear that not all of the large hanging knees that bracket the beams to the inside of the hull could be got from the oak pieces shipped from Latvia. Those right-angled knees are usually sawn from the root boles of trees where the main roots branch from the trunk. The Latvian suppliers had done their best but the heavy earthmoving equipment necessary to grub out the root boles was simply not available deep in the oak forests. I had wondered whether they could be dug out by hand, but my first short walk on the snow-covered frozen soil of a Latvian forest had persuaded me to keep that suggestion to myself.

Fortunately there are a number of Western Australian species that provide excellent timber for hanging knees, and large mature trees quite often have to be felled around the suburbs of Perth. Cutting out those massive knees is an interesting experience for the shipwrights and apprentices and adds to one's respect for the shipwrights of old who did it all with axes and pit saws rather than chain saws.

At this stage (late July 1998) the basic hull is built. The forecastle and aftercastle are being built and the upper deck beams prepared. Much work remains to be done laying the decks, fitting the mast steps and preparing the masts, spars and rigging.

The hull is impressively solid and very beautiful. Even encased in the shed of the Lotteries *Duyfken* Village shipyard and encumbered with the staging, she looks keen to sail. As expected, the shape of the ship as built is different from the original line plans that I drafted and from the more carefully and neatly drawn lines plan that Adriaan de Jong produced when the planking approached the turn of the bilge.

Some of the ideas we proposed about the sequence and logic of plank-first construction have proved impractical but the basic technique works well and even today might be a particularly efficient way of building a single ship because relatively little infrastructure and heavy lifting gear is necessary.

Adriaan de Jong:

I SAT down beside Adriaan de Jong at his drawing board in the Maritime Museum and discussed the replica construction as he had seen it from the start. 'First, we had to set the main dimensions for the ship from the length of the keel, a complex problem for which there was little information', he said.

It was known that she was of about fifty to sixty tonnes, so that was the starting point. They had to decide on the volume for that, and from various calculations found an average for length, width and depth. It was finally decided that the ship would be about seventy Amsterdam feet (about twenty metres) long — the length between the perpendiculars from stem to stern.

They studied the seventeenth-century pictures of *Duyfken*, and these helped in such matters as the rake of the sternpost. More difficult was the shape of the stem: 'We had to change it many times until we came up with a sort of generalisation for it.

They had hoped to make the keel out of two pieces, but the timber available did not allow this, so it had to be made from three pieces. Then the stem and stern had to be set up. The timber for the stem was such that it did not exactly fit the drawing, so they had to adjust the stempost to the available timber.

The design of the stern came again from studying pictures, and also the width of the wing transom, relative to the maximum width of the ship. There was a ratio between them. It was not known whether the ship builders used particular formulas in the sixteenth and seventeenth centuries, but there were certain ratios in use. There was a consistency in the width of the wing transom and the maximum width in various pictures and paintings. 'And we also had a few of those early ship contracts. That particular ratio was very close, if you took the wing transom to be one and the maximum width of the ship was about two.'

The other big question was the shape of the main frame. There were early signs of what it was going to be, but a seventeenth-century Dutch shipwreck, the SO1, confirmed what the design team thought was the case. It was seen that the maximum width of the ship occurred near or slightly below the waterline. And then the tumblehome (narrowing toward the top) was fairly straight running up all the way to the deck level.

The bilge itself (the curve out of the bottom to the vertical side) presented two options. Some ships had a hard chine (angle) from the bottom to the bilge, others a curved bilge. 'We decided on a more curved bilge rather than a hard chine. So you get the plane of the bottom coming up fourteen degrees, making a smooth curve to the tumblehome. Because of shallow entrances to harbours, the Dutch tended to build them flatter and wider so they had less draught, but to avoid unnecessary wide decks, the tumblehome compensated and also increased rigidity of the hull.'

De Jong said the shape of the main frame was used to carry that shape further fore and aft.

Adriaan de Jong and his model of the ship's boat.

He explained the planking process: 'You cannot simply force a plank into a shape you might make today with steel plating. You can only shape a plank in certain ways; timber will only bend a certain amount. If you force it further it will break. You can carry certain shapes fore and aft in a particular easy curve. If you draw these curves beyond the limits you get into trouble with shipwrights because they can't make it that way'.

Measuring the limits, he took flexible wooden battens to his drawing board to carry those curves fore and aft in his drawings. He could imagine the battens as planks and set up the drawing for it.

In the lines plan the ship was to have a cross-section or station at each metre of length, and for every cross-section there was a line for the shape of that particular frame, but the construction drawings were made to show the shape of a frame at every two metres.

The shipwrights did not need more forward detail. They had started the planking in the traditional way, and measured every second cross section to check that their work was going the right way. If the drawing suggested more twists than a plank would take, that had to be accepted as the limit, the extreme of the shape. Luckily, all the planks fitted the desired shape.

Especially in the beginning, the planks in the stern came out fairly flat from the keel and then they had to twist up to the upright of the sternpost. That was almost a ninety-degree change in the run of the plank. The strake was going from fourteen degrees to ninety degrees upright. To make that twist, of course, they had to do it in the traditional fire-bending way. It seemed doubtful that this could be done, but the oak planks allowed it, whereas any other timber might have failed.

Duyfken: Stern Assembly

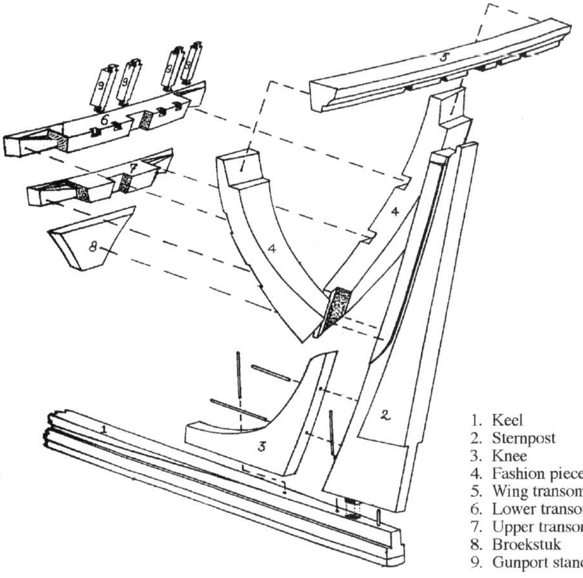

1. Keel
2. Sternpost
3. Knee
4. Fashion pieces
5. Wing transom
6. Lower transom
7. Upper transom
8. Broekstuk
9. Gunport stanchions

Drawn by Adriaan de Jong

De Jong concluded that in the old days the master shipwright would have known what the shape had to be, from doing it many times. But Bill Leonard and his team had not done it before. 'You tell us how it ought to be', he would say to the designers, 'and we'll try it out'. And that was how it was done!

The planking of the bottom of the ship was done in the traditional way with the planks clamped together, but the shape was developed guided by templates.

'When the bottom was developed with many little blocks (or cleats) holding the planks together, the shipwrights started to make a few floor timbers. They knew what their shape had to be, because the timbers had to fit the shape of the bottom. Once the floor timbers were laid, and with the planks up to the start of the turn of the bilge, a few first futtocks were set up in the ship as a guide for the bilge planks.

'After a few planks were set for the turn of the bilge, more of the first futtocks could be laid in place in the turn of the bilge. There was an overlap between the floor timber and the first futtock. So, step by step, the floor and the turn of the bilge were made that way — first the planks, then the floor timbers, then the turn of the bilge starting with some first futtocks, then the bilge planks and more first futtocks, making a rigid construction.

'At this point the planking had to go further up. Templates were used again to check what the shape should be (or had to be) and now they began setting up a number of framing timbers: the second futtocks, shaped according to the drawings and templates. These went much higher, and were fairly straight, set every two metres and held in place by a number of battens. This was a loose construction now, with all these futtocks standing up like the ribs of a skeleton, but the battens were there, with braces going across to maintain the shape.

'From that time, planking continued to the first wale. This first wale, a fair bit thicker than the planks, determined the sheer of the ship. "The beauty line", I like to call it.'

The planking was approaching that position, where the wale was going to be fitted to determine the sheer of the ship. At this stage the planking sequence differed slightly from the usual: planking was fitted up on to the second futtocks and a gap was left, then the sheer was determined fitting the first wale, and the last planks were made to fit in between.

Bill Leonard was keen, he noted, to make sure that all the planks gradually moved up to that sheer. You can't really say there is much difference in the changing shape of the planks to the sheer. The planks have a nice curve up to the sheer, matching the wale.

By now, along the side the first and second wales could be seen. There were to be four, the top one as the upper rail, which the shipwrights called by its Dutch name, the rahout. When required, the mast yard would rest on it. The wales were bevelled and rounded off to let water run away instead of sitting in pockets to cause rot.

'So we have the ends of the futtocks standing there and with the wale around it, you can already see what the shape of the ship is going to be.'

Shipwright Nick Truelove (left) fitting a heavy section of the rahout (rail) at the waist. John McLernon and Paul van Hooft are at the other end. On the right below, the template used.

The whole ship was yet to be further framed up from the inside. It was taking a long time to make all these timbers because each piece that was going to be part of the frame — floor timber, first or second futtock or others — needed to be shaped to fit its particular position inside the planks, and none of these lines were really straight.

In the middle of the ship the lines were easy enough, but coming to the bow the angles were increasingly difficult to realise. De Jong explained: 'In the bow itself some of the framing timbers are not set square to the keel, but are coming around, so they are rotating, so to speak'. These timbers were known as 'cant frames', meaning 'tilted, sloping, bevelled off'.

Another piece of the ship, one of the breast hooks, was already in place. These triangular heavy oak timbers fastened the ship together at the bow, tying the two sides together across the stem, with the apron timber making a continuous hook running up the stem, fairly wide at the top, and on each side of that the hawse pieces, to allow the hawse holes to be cut for the anchor ropes to go down through, port and starboard.

It could now be seen that our Little Dove would have a single, continuous deck running all the way fore-aft. De Jong likened it to the lid, the top of the vessel. The deck had to have hatches for cargo loading and for people to go down below, but otherwise it must be watertight.

In many of the old shipbuilding descriptions was the first deck, which Dutchmen call the 'overloop' (or overflow). That was also the water-shedding deck, so any water that came over, rainwater or waves breaking over it, had to run off to the lowest point, and at this lowest point would be the scuppers through which the water ran out into the ocean.

By this time the replica construction had reached the deck level. One could walk carefully over the curved deck beams laid across the ship, to view the ribbed formation below. Each oak beam was resting on the beam shelf, or deck clamp. The highest point in the deck could be seen in the middle and the lowest point on the sides, where the water would run into the scuppers.

At de Jong's drawing board he explained that although the deck beam was resting on the beam shelf, there was no direct joint between the beam and the side of the ship. But the deck beam was not just to support the deck; its primary main function was actually to hold the ship together, to join one side to the other.

De Jong noted that the other means of tying the deck beam to the side of the ship was the shipwrights' trick that Bill Leonard had spoken of earlier. In Witsen's text on shipbuilding and also in contemporary contracts, the deck beams were said to be joined to the beam shelf by means of dovetails. It was realised, however, that because of the tumblehome it would not be possible to lower a deck beam with some dovetails at each end vertically down in the receding dovetail recesses of the deck clamp AND have a proper joint.

Another method was therefore proposed. De Jong explained: 'In the first place one must realise that the deck beams are lightly curved so that where they rest on the deck clamp edge, this edge makes a small angle with the horizontal plane. Once the waterway is installed it will receive the spirketting on top of it. The spirketting does the reverse of what the clamp does: the clamp shoring up, the spirketting shoring down. Because of the tumblehome the lower edge of the spirketting, while being square to the ship's side, will make an angle with the horizontal plane.

'The latter angle however is the reverse of that of the deck clamp edge, together making the elusive dovetail. Considering further that the clamp and spirketting are bolted to the side of the ship, the deck beams are wedged in.'

Each of the fifteen deck beams had a knee at both ends, and there were also pairs for the three transoms at the stern. They all had to be individually shaped.

The waterways 300 millimetres wide running along the sides of the deck were important for the watertight connection of the deck to the ship. De Jong illustrated where the shipwrights would be making square openings in the waterways, and blocks about 200 millimetres square would be inserted. When the blocks were set in place in the construction, holes of at least eighty millimetres diameter would be drilled in them to get rid of the water. Four of these scuppers on each side at the lowest part of the sheer of the deck would shed the water.

There would be two more scuppers, but they were related to the two pumps which would bring the water up from the lowest point at the keel with a slightly different method of getting rid of the water. A channel would be cut into the deck beam, and the pump would shed its water directly into that, to run all the way to the side of the ship. Then, to send the water away through the side without breaching the watertight integrity, a lead liner would be laid in the channel in the beam to prevent rot, and this would become a lead pipe to discharge the water through the side of the ship.Generally the pumps would be manned only in times of trouble. In use, the ship might get some water below, but the pumps could each shift about 1.5 litres for every stroke, or about twenty litres per minute.

With the deck beams in place, the shipwrights were (in July 1998) cutting out all recesses to take the waterways.

They had almost built the rudder and would be making patterns for the pintles and gudgeons — in bronze, rather than the original wrought iron. At the Maritime Museum were a number of bronze pintles and gudgeons from excavated historic ships' rudders, and their dimensions were a guide for our ship. From top to bottom, *Duyfken*'s rudder would be almost five metres. It was very narrow — about seventy-five centimetres across at its widest — and fairly thick. But if the ship and her sails were properly balanced it should be quite effective, de Jong thought. That balance was to be found during sea trials.

The foremast was to be stepped on an extension of the keelson, like a shallow knee-shaped timber. Across the angled face of this timber the mast step would sit such that the foremast which rakes six degrees forward could be stepped squarely on it. The mainmast would be raked two degrees aft, and the mizzen mast about four degrees aft.

Although the ship was now more than two-thirds completed, there was still a lot of small work involved: the pumps to be made, the windlass, the bitts for the jeers (for lifting the lower yards of main and foremast), the bitts going all the way down to the hold. The beakhead construction was still to be started.

Finally, construction of the upper stern section had begun. Shipwrights were making up the timbers from the wing transom up to the deck transom and working their way up on the inside of the stern to the quarterdeck level.

De Jong indicated where there was provision for bunks for the crew, in the stern at the forward side of the quarter deck, in the forecastle, or below deck. Probably one-third of the crew would have been at rest while the others were on watch.

The skipper's 'hut' would be in the stern, the forward part of which they called the cabin, which was actually an undercover work area.

The helm was not a wheel but a combination of a tiller and whipstaff. The man at the helm would look at the luff of the sail if steering on the wind, or at the compass if steering a set course. The compass was in a binnacle or something similar where the man at the helm could see it to follow the given course. If there was a change of course ordered, he would be instructed to change, so many degrees north, south, east, west, and he would respond by shifting the helm, and would call out when he had reached the particular compass course.

John McLernon atop the stern, with its templates showing optional stern window shapes.

The ship was already 'pierced', as they used to say, for ten guns, four gunports each side of the deck and two, lower, in the stern. But there was more to the defence of the ship. De Jong recalled a few building contracts for 'vlieboten' (flyboats) which included small portholes in the aftercastle for the muskets to shoot through.

The documented story of the first *Duyfken* at Bantam, Java, in 1596, when she was stuck on a sandbank and almost captured while chasing an enemy junk, seemed to him to support this feature. While she was being extricated from the trap, some locals had rowed their prahus furiously to the ship and threatened to overrun her but apparently were repelled by volleys of musket fire as well as cannon blasts. De Jong wanted to get those musket portholes into the replica. As for the cannon pieces, the design team was thinking of having three-pounder cannon.

Also, on the quarterdeck, not a real deck in the sense of being watertight, there would be a 'bouchenet' or 'boevenet' — a grill through which the defenders would poke pikes to oppose enemy boarders.

Wood carver Jenny Scrayen and the dove emblem for the upper position on the stern. Her coat of arms of the City of Fremantle will fit below it.

I was especially interested in the ship's boat, having undertaken two research expeditions on Cape York Peninsula and in Torres Strait where *Duyfken*'s landing boat had obviously been used in 1606 to explore the rivers and parts of the coast, and presumably to carry fresh water and firewood out to the ship.

Adriaan de Jong thought *Duyfken*'s boat would have been about five metres long, that she probably carried about six people. I thought she would have been towed some of the time, such as between the main rivers, when the ship was surveying the coast of Cape York Peninsula.

One of the problems with the two-deck design of a small ship, as originally contemplated for the replica, had been stowing the ship's boat. Sources indicated that around *Duyfken*'s time, a seagoing ship would stow her boat below the upper deck on a long voyage such as sailing down the Atlantic and across the Indian Ocean.

De Jong had initially drawn up plans for the boat with an area where it could be placed below *Duyfken*'s upper deck. The layout was difficult, but it could be done — it only just fitted the available space. However the accepted alternative of a single deck in the waist of the ship provided for the boat to be stowed on the deck.

It could now be redesigned wider and more stable, a desirable improvement in my opinion, since we hoped she would be sailing about seven kilometres ashore to the Pennefather River when the time came, probably about April 2000, for a celebration of the historic 1606 discovery of *Duyfken*. In the wet season the waters there could be hazardous.

De Jong redrew plans for a typical ship's boat for the Little Dove: it was to be 5.1 metres long (eighteen Amsterdam feet), with a shallow draught of about 255–305 millimetres, suitable for the task of surveying an unknown coast and the rivers while *Duyfken* remained safely several miles out. On long voyages it would have been stowed on deck, in the waist between the mainmast and the forecastle.

The boat had a variety of uses and was essential for the safety of the ship. Sometimes it would sail ahead to sound the depths and guide the ship through difficult passages, and where the ship was to be moored in a spot where two anchors were required to prevent her swinging across into shallows, the boat would carry the anchors out and drop them at the correct positions.

A windlass was fitted to haul up the heavy anchors and lift water casks or other heavy items. Sometimes the boat would carry an anchor out to a safe point to which the ship could then be warped — as illustrated by the event in 1596 when the first *Duyfken* was kedged away from a shallow bank off Bantam while repelling Javanese boarders.[53]

De Jong built a 1:10 scale model of the planned boat. Originally he had intended only to develop the design of the planking, but the project captured him, so he completed it with sails, six oars, a leeboard each side serving like a keel or daggerboard to be lowered to reduce the tendency to make leeway — to slide sideways to leeward when sailing to windward — and a windlass across the middle of the boat.

The extremely lifelike model with its pleasing ebony finish to the clinker-built hull was placed on exhibition in a glass case in the *Duyfken* Ship Shop. Who would have guessed that the innovative De Jong had constructed the boat from balsa wood and cardboard coated with diluted tar, and with sails of cotton and nylon?

The full-sized ship's boat was to be built of oak, and Adriaan de Jong advised that for authenticity of the traditional seventeenth-century Dutch flat-bottom design it should be a fairly crude construction, hewn by axe and not too smoothly. He found that the manner of planking was still being used early in this century in fishing boats of the Hengst and Hoogaars class.

The Stone Cathedral

IT WAS the challenge contained in an ABC Radio National program that I heard by chance in 1993 which led me to begin the search for the missing details of the story of the *Duyfken*'s discovery of Australia. 'The Dutch Connection', a six-part series about the seventeenth-century Netherlanders, jointly produced by the Royal Netherlands Broadcasting Service and the Australian Broadcasting Corporation, won an international excellence award in 1990. In the program Cornelis de Heer spoke about early Dutch maritime history.

The discovery event was not itself news to me. Many years ago, at high school, I had read the textbook *Australia Since 1606*, by the South Australian historian G. V. Portus[54] with a brief account of some landings on Cape York Peninsula from the ship *Duyfken*. I remembered this author's comment: 'They called the natives wild cruel black savages because some of the crew who landed were murdered'.

In my school days nobody seemed to consider whether the Aborigines might have been defending their land from a foreign invasion, and in any case, the history of Australia at that time only appeared to begin seriously when the British came here to dump their unwanted convicts at Sydney Cove.

What arrested me in the ABC Radio program was the reference to an Aboriginal version of the events, a story said to have been handed down by generations of elders over four centuries. So, if *Duyfken* carried Australia's first European discoverers, this would be the oldest extant version of the events. Even if it was unwritten, it was a treasure that deserved a place in Australian history.

In the part of the broadcast that focused on the *Duyfken* landings, the producer Bill Bunbury briefly discussed the events of 1606 with a sociologist named Jan Roberts. She told him she had been working with the people of Mapoon on Cape York Peninsula in the 1970s, in negotiations with an offshore mining company. She described her experience at one of the Mapoon community meetings:

I was talking to the whole community gathered under the trees in the centre of the village. They asked me where this leading mining company, Shell, came from. I said it was mostly Dutch, with some English.

A man stood up, about fifty years old, and said: 'Huh — my grandfather and my father-in-law, they defeated the Dutch. We'll do it again'. Everyone applauded him.

... They told me the story of how the Dutch had landed at Cape Keerweer, immediately south of Aurukun. Their story was that the Dutch put up huts there and put in a well, and they made the Aborigines work, work, work, and they took their women.

And maybe it was a misunderstanding, maybe they were getting the gun out to go hunting. But the Aborigines thought they were going to kill them and so they got their spears. It came to a fight, and they came down to the beach in their canoes and set fire to the Dutch boat with fire sticks on the beach. The Dutch fled. The Aborigines said: 'Jan, we'll take you back down there and show you exactly where the Dutch are buried, eight Dutchmen'.

I had to consider whether it might be merely a colourful myth, influenced by the experience of nineteenth-century European colonisation of their lands, or if it could actually be unwritten history handed down through nearly four centuries, perhaps as a mixture of facts and myths. But fantasy or otherwise, that much of the story had captured me, and it seemed appropriate, after writing about the Dutch on the western coast of Australia and Australia's first shipwreck (the English *Tryall*, in 1622) to look back even further, to the story of *Duyfken*.

In 1644 the Dutch East India Company's council in Batavia wisely concluded, when discussing that remote land in the south yet to be identified as Australia, that: 'He who makes it his business to find out what the land produces must walk over it'. The wisdom of this approach appealed to my exploring instincts, but where would one look for the Aboriginal elders who were custodians of the strange story of the first Dutchmen who came to the Australian coast? They were not named in the radio program, and even if I could locate them, would they agree to tell me their story?

I reasoned that the search for the Aboriginal story would itself lead to a better understanding of the geography and history of the area and indicate the most likely landing places of the Dutch sailors 1606. But there was a good deal of initial research and planning to be done before it would be worthwhile heading for Cape York Peninsula.

By extreme good luck rather than skill, I found an appropriate contact and adviser, Ross Gardner, a civilian employee of the Royal Australian Navy at the HMAS *Cairns* North Queensland base. Ross was referred to me by his father Alan Gardner of Brisbane, one of the Australian soldiers about whom I had written in a Second World War book describing the conflict between the Australians and the Japanese at Milne Bay, Papua New Guinea in 1942.[55]

I learned that Ross Gardner had an impressive list of achievements, beginning with training as a motor mechanic, then qualifications and experience as a ship's master and ship's engineer, a stint as a national fisheries officer and a Certificate in Public Administration (Naval Staff College, 1992). Since 1989 he had been based at HMAS *Cairns* as the civilian manager of the Ordering Authority. Ross had spent three years on Yorke Island in Torres Strait as a Queensland fisheries officer, managing an enterprise development to assist the island community to progress from subsistence and traditional to commercial fishing; he had developed skills of communication with the Islanders. As a district officer with the Department of Harbours and Marine at Cairns from 1980 to 1984 he had been responsible for the operations of a team of eleven men involved in surveillance by land, sea and air, in Torres Strait and the Gulf of Carpentaria. I could not have found a more appropriate adviser and, even better, he became so enthusiastic about the proposed expedition that he volunteered to take some accumulated leave and join me. In September 1995 we began working jointly on the planning.

In Perth, Sir Ronald Wilson, former High Court justice, at that time chairman of the Anti-discrimination and Reconciliation Commission, seemed the key to my acceptance by the Aborigines. Our acquaintance went back a couple of decades to when I was a newspaper journalist dealing with lawyers, law courts and associated matters and he was a senior officer of the Western Australian Crown Law Department. As a High Court judge, and in his subsequent career, he remained the modest, courteous person I remembered.

Sir Ronald agreed to put his name on letters of introduction commending me to various individuals and leaders he knew in Aboriginal communities of Cape York Peninsula. I followed up these letters by telephoning and faxing the relevant Aboriginal offices, and was impressed by the friendly response, and surprised by the efficient manner and cooperation from places I had imagined to be something like bush camps in the remote northern corner of Australia.

Ross Gardner found the names of the two elders most likely to hold the traditional stories, Francis and Clive Yunkaporta. Their homeland was the southern area around Cape Keerweer, where something had caused *Duyfken*'s skipper Willem Jansz to turn back from his voyage in 1606. But it might be a difficult task getting to Keerweer and finding them — apparently you couldn't drive or fly there.

Clive Yunkaporta (left), Ross Gardner and Francis Yunkaporta in the dinghy, south of Cape Keerweer.

We agreed on a plan to fly from Cairns to Weipa, to charter a boat from there to take us about 130 kilometres south to Cape Keerweer and then to hire a four-wheel-drive vehicle to travel about seventy kilometres up from Weipa (as the crow flies) to Mapoon and the Wenlock River, named as the Batavia River in several documents referring to a fatal Dutch encounter there with Aborigines. In Perth and Holland my interested friends at Fugro Survey accepted my budget for the costs of the expedition.

A Cairns friend of Ross Gardner, Michael Fordham, then a regional manager of the Queensland Government Department of Housing, Local Government and Planning, was also keen to join in the expedition. Mr Fordham's region included the western coast of Cape York Peninsula, he was a lieutenant commander in the naval reserve and, like Ross, an experienced bushman who could understand and get along well with the Aborigines.

Inquiries about a suitable charter boat at Weipa — one of shallow draught but capable of carrying at least six people — led Ross Gardner to Alan Otto, who ran a ten-metre catamaran, the *Jamie Lee*, mainly for eight-day barramundi fishing trips.

Ross Gardner, Michael Fordham and I flew to Weipa, site of the world's largest bauxite mine, operated by the giant mining company Comalco, and at the airport we hired a four-wheel-drive vehicle. We were scheduled to join our charter boat next morning at the dock known as Evans Landing, and accommodation was scarce, but Michael Fordham soon organised some for us overnight at the Comalco quarters. The eye-catching red cliffs and soils of Cape York Peninsula had been noted long ago, but in 1956 geologist Hay Evans discovered the potential of the massive bauxite deposits north and south of Weipa. The land around the Weipa mission station was excised from the Aboriginal reserve and made available to Comalco, and in 1960 the company built a new sawmill at Aurukun and drilled the first permanent water bore.

Next morning, at Evans Landing, we found the *Jamie Lee* waiting for us and met the owner-skipper Alan Otto and his deck hand 'Crazy Pete', the star entertainer of Weipa's Albatross Hotel. In spite of his sobriquet, Pete showed himself quite sane, efficient at his duties on the boat, and both he and Alan Otto were interested in the aims of our expedition.

Heading down the warm pale brown stream of the Embley River into the wide Albatross Bay, named Vliege (Fly) Bay on Willem Jansz' chart, we discussed the plans. First call would be at Aurukun, the Aboriginal settlement on the Archer River about eighty kilometres south, where we would pick up the two Yunkaporta brothers and take them with us sixty-eight kilometres south to Cape Keerweer, which was their traditional land. By this time the question of finding these men somewhere around Aurukun had been sorted out with the Aboriginal community, the local authority and the police. Everybody was on our side, it seemed.

It was a smooth trip down the coast and we came cautiously around the shallows of Worbody Point into the entrance of Archer River, which was aptly named Dubbel Revier on the *Duyfken* chart, being joined by the Ward, Watson and West Archer rivers.

Our boat anchored about 150 metres from a sandy beach where a four-wheel-drive vehicle with police markings was waiting. Alan Otto made radio contact with the shore, and soon a police officer was heading out to us in a dinghy. Ashore, the police vehicle transported us up the track to the police station for a friendly discussion with Sergeant Jim Rankin, and then to the Aurukun Shire Council office to talk with the mayor Jacob Wolmby, a senior member of an important family in the Aboriginal community. The Yunkaporta brothers were out of town, but Mr Wolmby and his brother Silas told us not to worry: they were happy about the arrangements, and would be ready to join us in the morning.

From on board the *Jamie Lee* early next morning we saw the police vehicle arrive at the beach, and the police dinghy brought two elderly Aboriginal men out to us. The police officer cheerfully introduced Clive and Francis Yunkaporta to us, they thanked him quietly, we all shook hands, and the officer returned to shore alone.

The barefooted brothers were dressed in shorts and shirt and both were agile and alert. Francis, the younger, who was in his seventies, had his left foot bandaged. He was well built, with barrel chest and thick muscled arms. His skin shone smooth dark bronze and the hair above his broad friendly face was a shock of steel grey. His moustache and beard were neatly trimmed. Clive was a smaller, slender man with dark brown curly hair and smaller beard. His clouded eyes betrayed his age; I estimated him as about a decade older than Francis.

We headed slowly down the river, passing green-grey stretches of mangroves and low trees, Alan Otto and deckie Pete carefully checking the pale brown shallow patches to avoid grounding in the low tide. I told the two Aborigines of my interest in the story of the Dutch visit in 1606 and the assertion of historians that some of the Dutch sailors died in encounters with the Aboriginal people of Cape York Peninsula, that I wanted to learn and understand the traditional story which I was told had been handed down to them by their ancestors.

Clive and Francis said they were looking forward to going back to visit their birthplace and home country at Keerweer, and they were prepared to tell me the traditional stories and also show me the places, so that I could write the true story for all Australians. Initially I had difficulty in understanding the Yunkaportas, but they seemed to understand most of what I was saying. Obviously the weakness was on my side: the elders were addressing me in English, a language other than their own. With familiarity I became more confident, while the elders seemed to grow in stature as the wise men who understood the mysteries of this country, its creation and the ancient natural laws and religious customs. Ross Gardner had advised me not to rush things, to allow the stories to emerge at the appropriate times.

Francis shaded his eyes against the bright early morning sun as he looked fondly astern at the land around the Archer River bank: 'My people planted all those trees and plants you see'. I had no doubt that white Australians generally would have dismissed this thirsty-looking vegetation as worthless 'scrub'. A little later he was giving me a cryptic description of Clive Yunkaporta: 'My older brother Clive, when he write stories, a little bit longer, but when he speak, a little bit shorter. He write stories for young ones to read too'.

A few coconut palms waved pleasantly over the yellow sand spit at Worbody Point beyond Wallaby Island, and a big seaward sprawl of mangroves marked the southern exit of the Archer River. It was a careful course for several shallow miles out to sea before we could turn south for Cape Keerweer, sixty-eight unsurveyed kilometres down, as my June 1992 naval chart indicated.

Something ashore caught my eye briefly; a slender dark column of smoke snaking up above the tree line, back toward Aurukun. Then my interest was captured by the pretty pattern, black, grey, pink, and white, of the bauxite cliffs between the white sandy beach and the distant dark blue horizon of low trees, stretching south all the same as far as the eye could follow. The crew of *Duyfken* might well have found this monotonous.

At 7.15 a.m. Francis pointed to the shore about six kilometres away: 'Half-way'. I consulted my chart and then asked Alan Otto, a navigator who knew this coast well from about twenty years' experience in the Gulf. 'He's bloody right you know', he replied. The *Jamie Lee* cruised on for a while, and then turned in close to the shore. There was a shallow bank to seaward.

When another slender column of smoke rose above the trees and hovered in the stillness a few miles inland I saw that the elders were also quietly observing it, and asked: 'Is that another smoke signal?'. They nodded in affirmation.

'A message to your people at Keerweer?'. 'Oh yes', was the casual reply, and I took it that this was quite ordinary, just an illustration of the system of communication the elders would expect to continue.

About forty-five minutes later Francis indicated a line of trees a few kilometres inland along a river. This was Peret outstation, one of the Aboriginal cattle stations on the Cape York Peninsula's coastal Aboriginal Trust Area. The map showed a track ending here beside an airstrip, and it wound far down past Tiwi outstation and another airstrip about thirty-five kilometres south-east of Cape Keerweer, then to the Kendall and Holroyd rivers. Soon another thin column of dark smoke rose from a few kilometres inland and hung in the sky. Now we were cruising in smooth, amber water closer to shore. Alan Otto pointed to the remains of a dugout canoe, about twelve metres long. From New Guinea, it had been washed ashore in a mighty cyclone some years ago, and each year since then the storms had carried it farther up the beach.

Pete the deckie took the helm while Francis gave the skipper some advice about a sandbank in the Kirke River at Cape Keerweer. I didn't know where we would be heading when we got there; it would depend on the what the Aborigines wanted to show us, but Alan Otto obviously had it all in hand.

At 9.00 a.m., when we were about fifteen kilometres north of the Kirke River, Francis pointed ashore. 'Another smoke signal?', I asked. 'Yes', he agreed calmly. 'And another one further down — they've started going to meet us.' Who 'they' were, I had no idea, but the ancient means of communication seemed to be working well. Later the elders were at the rail keenly examining the shore line. 'The oak trees', they announced, and explained that this grove of greener trees identified the nearby site on the beach where the Dutch sailors had dug wells for water in 1606. There had been a battle between them and their ancestors, the Aboriginal warriors, and some Dutch bodies were buried here.

I would have liked the boat to come ashore so I could walk up the beach to the relevant places, but the elders told me not to worry, we would be able to get there easily from the Kirke River. Alan Otto agreed, and with the sun now blazing on our voyage it was time for a cold beer all round.

Cape Keerweer, known by the Aborigine people as Thewena, is shown on the marine chart as a small black dot at the latitude of 13°55.5′S, and is not easy to find from sea. The narrow mouth of the Kirke River lies just above it and the slim river snakes northward, then through a series of crazy loops until, about ten kilometres above the mouth it reaches what is more like a broad lake extending about ten kilometres down to another smaller lake, and more of the river disappearing to the south-east. Very cautiously we moved past a point with beautiful golden beach overhung with green trees and a few coconut palms. Inside the entrance the river came down from the north.

We watched an aluminium dinghy streak in, about six metres long, with an outboard and carrying about half-a-dozen young Aborigines. As they jumped out and confidently hauled the dinghy up the beach I noticed among them a mother and wide-eyed brown baby, obviously enjoying the activity. I learned from Mike Fordham, who chatted with them, that they were a family who had come down from Aurukun to visit relatives at Keerweer. It was not uncommon for such groups to make voyages of about 150 or 200 kilometres along this lonely coast.

Now I learned then that we had another intriguing question to follow up, south of Cape Keerweer, before sailing up the Kirke River. The elders were keen for us first to visit a very important sacred site. 'We would like to take you to the moving stone', Clive Yunkaporta said precisely, and Francis explained it briefly for me:

> Way back in the beginning, in the dreamtime, our ancestors passed this story on to our people ...
>
> There were two girls who belonged to the people from the coast country about one or two miles south of the Kirke River. This southern area of Cape Keerweer belongs to the Yunkaporta clan.
>
> The girls, who were sisters, decided to test their power against the power of a sacred and powerful rock which was lying on the ground in their country. This rock was magical and mysterious. It could shift along the ground by itself and was known to the people as the moving stone.
>
> One day the two sisters grabbed the stone and together they lifted it off the ground. Then with their own power they flew up in the sky, holding the stone between them, and held it up two or three hundred metres in the sky.
>
> But before long, the power of the stone started to overcome them. Their arms became paralysed and they were forced to let go the stone. It dropped back to the ground in the area where it lies today, not far from the beach.

The stone still has power. It is able to move of its own accord. Francis said his grandfather was one of very few people who had seen the stone move. 'It moves along the beach — backward and forward — but we can't see the stone move because it's secret.'

It was given as a gift to his generation, Francis said, passed on by their forefathers, protected by them so that no one could take it away. One day in 1940 the superintendent of the Presbyterian mission at Aurukun, Reverend Bill MacKenzie, went down to Cape Keerweer with the traditional elders to move the stone away from the tribal land. Evidently the mission superintendent resented this symbol of traditional Aboriginal beliefs.

When they got there they put a rope around the stone and dragged it a few yards, and the rope snapped. MacKenzie changed his mind and got a wood chisel and tried to take a piece of the stone back to Aurukun. When my father learned what McKenzie had done he was very angry and he told them to get away and leave the stone where it was.

Later some other Keerweer people added, with obvious satisfaction, that Reverend McKenzie had been so furious when he could not chisel a piece off the stone that he kicked it and injured his leg.

Francis told me there was another Keerweer story I should know, the legend of the Quail:

Our people did not believe in the Quail. They did not realise that the Quail was given to them as a source of power and they did not want to put their faith in him. But a few did believe.

One day they all went down to the river of Cape Keerweer for fishing, and when they came back the boy Quail was dead.

Some of them mourned, but big mobs did not, so the father Quail was upset and angry. He took off and flew around the mountain. The sea felt his anger, and big waves came in over the land. The Quail then picked up the mountain and everything that went with it, and took it inland and placed it near the town of Coen [194 kilometres east of Keerweer] ...

Our skipper Alan Otto advised that the water south of Cape Keerweer was too shallow for us to risk taking the *Jamie Lee* down on a search for the moving stone, but he thought we might be able to borrow a dinghy from Eddie Fisher's Landing, a short distance up the Kirke River. Soon we saw a little settlement on the left bank, a landing and Eddie Fisher's barge moored there. We pulled in alongside, and Alan and Ross clambered aboard the bulky vessel, returning in a few minutes with the news that Eddie's boat was available.

We piled into this dinghy and set off smoothly down the river. The Yunkaporta brothers sat on the bow seat and I was kneeling just behind them, anxious to hear, above the row of the outboard motor, what they might say about finding the mysterious stone for us. The others sat in the stern.

Out in the shallow pale brown water of the open sea, heading south, the dinghy was well balanced and riding easily, so we increased speed. Past what my chart had shown[56] as Cape Keerweer — even if it barely resembled the dictionary definition of a cape as 'a headland or promontory' — the southerly wind picked up and we bumped and skipped over the little waves. Clive Yunkaporta raised his right arm. 'Slow, slow', he called. I relayed the advice to the stern, the outboard noise diminished and we reduced speed, wondering about the reason for it.

Clive leaned over the seaward side and trailed his right arm in the water while he addressed the sea and the sky in his own language for a couple of minutes. Then he turned back to explain the situation to me in my language: 'I was telling the wind and the water to be calm', he said. 'I told them I was bringing my friends to hear the stories of my people and to see the stone. Now it is calmer.'

A couple of miles farther down the low-lying coast both elders appeared excited. 'In here', they said. We headed toward the beach. 'No: down further.' We continued slowly southward. I noticed Ross in the stern, holding out his hands, palms up, in silent expression of the doubts I too was beginning to feel about the mysterious quest.

The aging brothers intently studied the dry stunted bush land just beyond the blinding white beach sand. 'Here', said Francis. Clive shook his head. 'It should be farther down', he said, but: 'Okay, we go in here and have a look'.

About 100 metres out, we were motoring slowly shoreward. The elders were barefoot, and we decided to leave our boots in the boat; a mistake, we would find. At fifty metres out, we were drifting in little more than a metre of clear water, and it seemed the appropriate time to hop out to haul the dinghy ashore. I stood up and had a leg half over the side when I felt an arresting hand on my arm. 'Not yet', said Francis, 'too many crocs.' I took his word, and waited for a while. Later I would discover that his warning was founded on his own experience.

While I was hauling the boat up to the beach Ross and Michael were carrying Francis ashore. His sore foot was giving him trouble walking. He decided to wait on the beach while his older brother Clive, who still insisted that the stone was farther south, would lead our search for it. He gazed inland, then down the coast, and set off, surprisingly fast and agile for his eighty-plus years, in his chosen middle direction, about south-east.

We were soon running to keep up with his fast walk. Then we were skipping and jumping because the ground beyond the damp beach sand was hot as hell. I tried landing my burning feet in Clive's untroubled footprints, but the heat was still painful so I began hopping on to clumps of spiky spinifex — it was less painful than on open ground. The others, although well experienced in cross-country travelling, were also having problems.

Clive paused while we all caught up with him and he indicated a dark solid object about fifty metres farther. We went on together — there were more plants and some small trees about and the ground seemed just slightly cooler — and here was the stone, a fairly symmetrical cube of nearly one metre of dark grey basalt, sitting firmly, its tonne or more of weight sunk partly into the ground.

We stood still for a minute or two, silently wondering how — other than from 300 metres above in the faltering hands of the legendary sisters of the Dreamtime — the heavy block could have got to this remote wilderness. And from where? It seemed quite alien to the geology of Cape Keerweer or anything we had seen all the way down from Weipa. Why would anybody want to take it there? And who? ... a student of Henry Moore? But this was not the time to dwell on such questions. In the white man's holy places, would people interrupt a priest, chanting in Latin, to ask about matters of science and logic?

The stone might be seen as the cathedral of Cape Keerweer's elder son Clive Yunkaporta. The wise man had brought us to the holy place to reveal the truths to us, and the service was about to begin. He sat on the ground, put a hand under his left arm and then placed it on the stone. Now his clouded ancient eyes looked through us into the distance and he began talking in his language, chanting, singing to the stone and the spirits. Then he put his right hand under his left armpit. There was silence. He looked down and explained.

He had begun by putting his smell, from under his arm, on the stone; that was the law, the customary way for him to greet the stone. He had told the stone he had brought Leaning Tree and these other friends of his people to hear the stories about this place, about his people and those things of long ago. 'Leaning Tree' proved to be me, although I never found out how the name was derived.

It was a spellbinding experience. In the silence the old eyes turned toward me and I understood. I asked: 'May I touch the stone?'.

'Yes, but wait.' As I knelt beside the stone he put his hand under his left armpit and then on to my knee. 'Now you can touch it.'

I placed my hand on the stone, and as the chanting began again I thought of Clive Yunkaporta's people who had come to Cape York Peninsula at least 40,000 years ago, perhaps 100,000 years, long before Genesis, Noah and his Ark, Moses and his stone tablets, and Jonah swallowed by a whale in the teachings of the fundamentalist church of my father, long dead now, his red brick church demolished for a hotel parking lot.

Ross Gardner, also anointed by the elder's appropriate smell, replaced me beside the stone. While his hand was resting on the top surface of it he noticed where somebody had attempted to chip a small piece off the side. 'Yes', Clive noted, 'that's where Reverend MacKenzie tried to chip off a piece'.

Ross asked, lyrically: 'Can Michael touch 'im, try for lift 'im, not shift 'im but see if 'e can lift 'im?'.

'No, you can't', was the polite but firm reply. However Michael was allowed to touch the stone. Clive explained that if anybody tried to lift the stone a great charge of electricity would paralyse their arms. When some people had tried to move the stone a big wind had come, a great whistling noise, so people knew they must not try to move or touch it. The stone had moved of its own accord since he had last seen it a year ago. 'It was over there', he indicated a nearby gully, 'and now it's gone back to a forbidden place. Not far.' In the last year the sea had been all over this area, Clive said, and it looked that way to me too.

Now, in this remote wilderness, the stone was at rest again, just a lone, inexplicable basalt cube, and the elder's powerful spiritual presence had diminished to a level of quiet dignity. We departed the scene, suffering again the hellish heat of about 100 metres of the burning ground and beach sand, all but the biblical burning bush, and got back into the dinghy to sail back northward to the *Jamie Lee* at Eddie Fisher's Landing.

Back on our boat, Ross Gardner told me he had heard a non-Aboriginal explanation that the stone could have come to Australia as ballast or an anchor of a Dutch or English ship centuries ago, and he wondered if I had seen anything like it on the Dutch shipwrecks off the Western Australian coast. I had indeed seen and brought up from the depths some of the little Dutch building bricks carried as ballast on one seventeenth-century wreck, and had discussed with maritime archaeologists the small stones used as ballast on others. But that explanation did not hold for me, I said. This stone would have been far too heavy for the sailors to handle, let alone to bring ashore in a ship's boat. Later we debated

Michael Fordham and the elder Clive Yunkaporta, hands on the sacred moving stone.

whether from time to time down the centuries mighty cyclones had swept over this area, shifting the dunes back and forth, changing coastlines and river courses, covering and revealing, perhaps even moving the stone. But the origin didn't really matter, we decided. It remained a powerful symbol of endurance for the Wik people of Aurukun on the western side of Cape York Peninsula, down to Cape Keerweer and immediately south, people who speak the Wik-Ngathan language.

I noticed a crooked scar, about twenty centimetres long, running across Francis Yunkaporta's left thigh, and asked him how he got it. 'A croc', he said. Some years ago at Keerweer, he had been in the sea fishing with a spear not far from shore when he felt a bump and something dragging, seizing his leg. Quite suddenly the water around him turned red. He looked down through the turbulence and made out the ugly head of a crocodile, so he punched his fist with his all his strength at one of its eyes. Then there was more blood spreading up in the water — this time from the crocodile, he thought, because it let go and he was able to stagger ashore to safety. I saw from the size of the scar that it must have been a massive wound. After escaping the saltwater crocodile's jaws Francis was fortunate not to have died from loss of blood or blood poisoning.

Clive Yunkaporta expressed his great satisfaction about our visit to the moving stone. The old people and the friends would be arriving soon, he announced — apparently summoned in the traditional way by the smoke signals that had followed us down from Aurukun.

Now back at our anchorage in the Kirke River, Francis Yunkaporta was prepared to reveal to me his traditional version of the Dutch landings of 1606. I could question him and his older brother later, if I wanted to, and the people would join us too. This was the proper way, he said.

The story had been told to him by some great warriors: his mother's father and her father's father, also his father and his father's uncle. Francis, his father's uncle and his people came from the south side of Cape Keerweer's Kirke River.

The Aboriginal people saw the first Dutch ship north of the mouth of this river in 1606. They saw a big mob of logs that were huge, very big with lots of devils on them. The devils looked strange. Their skin looked different and they were white. The Aboriginal people wanted to know why they had come.

The warriors were hiding and watching from the bush. They wanted to see what these strangers were going to do.

The devils came from the logs and rowed up to the beach. More than one boat came ashore.[57] They were standing on the beach and looking around, and holding a stick which took them to a place near the bank of the river. The strangers had shovels and began to dig.

The warriors saw what was going on and could not sit any longer. Some warriors went to the strangers to find out who they were and what they were doing on their land. This was not going to be easy because the language they spoke to each other was not the same.

Again, next morning, the warriors came back [to the beach] and the Dutchmen showed them how to use a pick, a shovel and an iron crowbar to dig the ground. The Aborigine people had always used a yam stick to dig and a bay shell to lift the dirt. The Dutch got the warriors to dig the well for them. The well took many days to dig. During that time, the warriors imagined the Dutch people came from the sky. They showed the warriors how to smoke tobacco, how to bake damper and how to boil tea. They also showed them a gun to shoot ducks.

But the warriors asked each other if all these things were good. When they tried smoking the tobacco they asked each other what they thought about it, was it good or bad? One man said it was no good, that the smoke which came out of their mouth was a spirit from their body or one spirit from the past. The warriors were angry. They saw the Dutch people as 'onya', meaning a nuisance that brought all this evil. The warriors went back to their camp.

The Dutch dug one well and a second and then they were digging another alongside the beach. At the end of each day the Dutch would go back to their ship.

One day the warriors were greatly upset because the Dutch people were misbehaving. This caused much trouble.

The warriors came back with anger to where the Dutch were working. For a while, the warriors stood there with the Dutch. Then the Dutch went back to work at the bottom of the well. The warriors went to the well and jumped in on to the men below by surprise and beat them. This was Aboriginal law.

There was much fighting between the Dutch and the warriors. The Dutch shot many Aboriginal people along the river and in the bush land. Also, the warriors speared and killed some Dutchmen and made the Dutch go back to their ship. The warriors and the Aboriginal people saw the Dutch return back to where they came from.

The area along the river was called Cape Keerweer by those first Dutch explorers. Today the Yunkaportas' people still use the name Cape Keerweer for that river country …

That evidently was the end of the formal version of the story, for the time being.

Next morning Francis was ready to add to and explain what he had earlier told us of the encounter at Keerweer in 1606. He was born in 1923 and his ancestry went back via his grandfather Peter Pumpkin and then long, long ago to the warrior leader who defeated the Dutch.[58]

'That old grandfather of ours [meaning of 1606], he tackled the Dutch. He was a leader of the army, you know, a great warrior. The old man he tell us ...'

Francis switched briefly to the present: 'A brother of mine, he got a station over here, three miles from Cape Keerweer'.

Back in the distant past: 'All this wetland country was flooded by the angry sea and many of our people were swept away and drowned. That was after the child died of the Quail. The mountain was taken away and the big wave came, long, long time ago. That's the history the old man [Francis' and Clive's grandfather] told us ... It was after that time that my ancestors saw the Dutch come'.

Where was the battle between the Aborigines and the Dutch? 'The place we be going later, and the well they [the Dutch] digged along the beach, it was the one we used to use. That was a good one, but it's covered up by sand now. But you may see some of the bones ... a lot of things come up. A lot of bones were buried by the old people ... Lots of our people killed, along the river bank and in the wetlands.'

How many? 'We don't know. Those years long ago, that time, we don't know how to count.' Francis held up his fingers to illustrate. 'We only know then how to say one two three. We don't know how to spell cat those days. All we be warriors, all of us then. With spears.'

Long after that, the missionary had come: 'That Reverend Bill MacKenzie, superintendent of the Presbyterian mission at Aurukun, he come down with police and trackers. Took 'em away my people in chains all the way along this road, maybe one night, two nights, three nights ...'.

I heard the sound of a tractor. It was towing a trailer crowded with families of the Keerweer outstations. 'We must talk with the people too', Francis said, and so we went ashore to sit down with them in the shade of a sort of open-fronted hall. The congregation included some old people, families and excited, chattering young children. He introduced us, explaining that we had come to Keerweer to hear the stories of the people.

We continued with the issue of his people taken away with chains around their necks and forced to leave their children with the missionary at Aurukun. There the children, including Francis, were locked in a building like a big gaol, and the parents were told to go back into the bush. The parents were not allowed to see their children more than once or twice a year. These things stopped about 1968. By then Reverend Bill MacKenzie was dead.

A man named Morrison stood up to say he was one of children taken to the mission. And a woman, Hazel, called out that she was too. There was a clamour of similar voices. 'Maybe all the ones you can see in town', Francis Yunkaporta said.

After some discussion about poor food at the mission, I turned to the very slender Morrison Wolmby and commented: 'Morrison didn't put on much weight from that mission tucker'. He replied with good humour: 'I be like skinny fella all the time'. The Reverend Bill MacKenzie seemed to be remembered with a mixture of resentment and some gratitude for the education of children although they were forcibly taken from their parents.

Francis raised his voice. 'Let us get away from that now.' He turned to me, but he was addressing the people too, and said slowly, very precisely: 'Your question was about the Dutch. You ask me how it started, with the warriors.

'Our great grandfather was a warrior, my mother's father. And we are descendants of that old Peter. That was name Dutch gave him.'

Francis Yunkaporta narrated that when some of long-time-ago Peter's people had seen the Dutch ship they thought was a big, big log they went and told the other people. Then some of the devils came ashore in a boat. The hidden warriors watched them come closer.

When the warriors found out these white people speak: 'Oh they can speak', they go back and tell the other Aboriginal people. Then they speak and very closely and make 'em friends. They showed him how to make damper, how to boil tea. Out in scrub. Showed Peter how to use a gun to shoot lot of ducks.

When he was showing the old man how to shoot ducks the old man put the gun barrel on his shoulder. The Dutchman said: 'No, no, no, not like that. You got to put this stock in the shoulder and fire like that, bang'.

'Woh', they said. 'It's the devil's sign.' It made them angry, made the Aboriginal people very angry.

Of course, Peter talk his own language, our language, and the Dutch speak their own language, like you speaking your own language now. I could be able speak my own language to you.

The Dutch can't hear what is happening. They work in signs, they do a miracle thing, you know, by signs, Dutch and Aboriginal people. You want to dig a hole, you go like that [he indicated]. Cut wood, there. Dig for water ...

How did the Dutch know there was water there? 'Well, they did, soon as they came, asking for water, making signs ...'

So your people gave them water? 'Oh no, no. Didn't give 'em any water at all. They making signs if they can get water in area around the place. They point down to the ground. Stick. Then they want to make the Aborigines to work, my group, to work with them.'

Did the Aborigines work? 'Well, they did. They started off the well. From there they did another well on this [western] side of the river. Where that rock wall, on this side. And that's where that battle started.' The description of the location would become clearer when we visited the site later in the day.

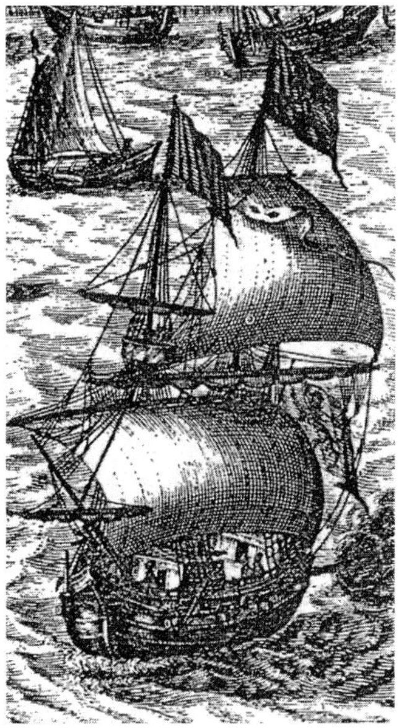

'Before that, they found that there were a couple of girls taken away by the Dutch. And that's where the argument started. They said, oh well, those girls never been found, they must be on the ship. This was not behaviour, you know, er, it was misbehaviour with those people.'

Francis was leading us and his now silent audience to the point of the violent climax. 'Two girls in boat when happening. So when they [the Dutch] coming back again to shore where they worked, my ancestor Peter, he said: "Look, when they come, we let 'em jump into the well and let 'em dig".

'Anyway, they all jumped in. And there was a couple of them down there with [a cask or barrel]. And a few more of the others jumped in. Then all of a sudden Peter jumped in and smothered them in that well ... about five or six, I think. My group, Peter and family hit 'em with sticks. Then, it could be a general or someone, an officer, he came along and he went back and brought another group.

'Yes, they start shooting then. Bang, bang. My ancestors crossing the river, and they shooting them. What you call that thing, they had to load 'im with the powder? — Yes, muskets. While they loading that, some got speared. Then bang, bang, knock about eight or nine of my mob. Up on the bank, children and all, poor little buggers. Cleaned them out. We lucky, our father, we escaped.'

Francis added, evidently to emphasise that he was referring to the early events long before English settlement: 'Captain Cook wasn't here. Oh yes, Captain Cook was not on this west coast; all we know is he was doin' the eastern coast'.

We see that well today? 'Yes. But I not go, I got sore foot. I lay in boat.'

I raised the delicate question about the 1606 report of Captain John Saris that nine of the *Duyfken* landing boat crew had been 'killed by the Heathens, which are man-eaters', and asked: 'Do you know about that story?'.

Francis Yunkaporta was not offended as he denied it. 'No. I never heard that, cannibals. My ancestors would have told the story, and they never told us that. Our ancestors told us they buried the Dutch bodies. No; the warriors did not eat the flesh of their enemies.'

Francis added that about twenty years ago a very wise man named Dr Peter Sutton had come from a South Australian university to Cape York Peninsula and learnt to speak the languages of the Wik people. The Aborigines had told Peter Sutton the stories in their own language. He was a very good man, they reiterated, and I should talk to him. The other Aborigines in this audience nodded; evidently they all remembered Dr Sutton with great respect. I said I would try to find him.

Francis Yunkaporta ended on what seemed an eminently sensible note: 'We keep telling our people in town about the time when the Dutch were here — my people in the town. Some people say: "What about the Dutch, forget it". Well, we won't say let's forget about the Dutch, but we might say we can forgive. Not forget, otherwise we won't carry on the Dutch history'.

Now there was an urgent problem for the present. A woman, relative of the Yunkaportas, was very sick and needed to go to hospital as soon as possible. The best way to get her there would be for an aircraft to fly down from Weipa or Aurukun and pick her up from the airstrip at Aayka. This outstation was about five kilometres across the salt coastal flats from Cape Keerweer as the crow flies. The Kirke River ran from about thirteen kilometres north, where its broad section reached down and narrowed about twenty kilometres south-east to Aayka and the airstrip at Ti-Tree Outstation. The main difficulty was that nobody was available who could work the radio at the outstation to organise for an aircraft.

We held a conference on the *Jamie Lee*. If necessary we could carry her on the boat, but it would take a couple of days to get back to Weipa. But Michael Fordham volunteered to solve the problem. He could work the radio at Aayka, and understood the procedure involved for arranging to get the aircraft for medical purposes. He could travel to Aayka with the Aborigines on the tractor and trailer and then be brought back to a rendezvous with us and the *Jamie Lee* somewhere up the Kirke River, at a suitable spot the Aborigines indicated to our skipper Alan Otto.

The tractor rumbled and bumped away on its cross-country journey and Michael Fordham waved precariously to us from the flat-top trailer. He was aware that the distance on the map might have little relevance to the course through the wetlands and bush — out there and back to us at a river landing.

We headed north up the river, narrow but reasonably straight for a while. Francis admired some coolabah and bloodwood trees and what would be termed as scrub in his beloved brown country. Once there had been plenty of scrub, he said nostalgically, but much of it had been taken away. 'My mother's country. Them trees, 'e used to supply us with food, y'know. Money can't buy. Trees that we can eat the fruit. Creation made that for the people to eat.'

What else did you eat from here? 'Lots of things, wallabies, snakes, goannas, bandicoots, birds, possums, fish — spear 'em. No lines back in those old days. Came lately. But they good. Europeans come along with the lines, and soldiers come.'[59]

After several miles the narrow Kirke River swung sharply to the east, then back west and north again and into a series of maddening loops in which the dense mangrove banks ahead kept converging, almost like tunnels. I guessed that monsoon floods had swept down here carrying vast quantities of mud and sand and tree branches to be dumped at the many sharp bends and beside mangrove islets where the stream was slowed.

The river widened again and we began to look out for the rendezvous point, said to be near the 'rock wall', whatever that might mean. The eastern side of the river was lined with hungry low, greyish, mangroves, but on the western (sea) side the tall green mangroves merged with vigorous trees — no indication yet of any sort of wall. 'Soon', the elders told us patiently, but the river took us around more of its crazy bends. Then, rising out of the western tree line several hundred metres ahead, several puffs of white smoke followed by a small but steady and gathering white cloud. 'Smoke signal', I said. The Yunkaporta brothers were strangely silent this time. Smoke, but not a signal, I gathered.

We rounded another couple of bends, and ahead out of the shadows of a gloomy tangle of trees rose the orange-red flames of a healthy little fire. Then we could make out the cheerful sight of Michael Fordham and four Aborigines standing nearby among the dry brown grass of a little ledge, their dark faces gleaming in the flashes from the fire.

Alan Otto brought the *Jamie Lee* in close enough for Michael to reach across from the embankment and clamber aboard the boat. His mission had been accomplished; he had radioed from Aayka Outstation for a plane to come and take the sick woman to hospital. He had lit the fire, partly to drive away the mosquitoes and flies.

We set off down the river for our significant destination, the area of Australia's first fatal conflict between white and black, according to the traditional Aboriginal version. Back in the winding course of confusing loops and islands in the stream it seemed strange that such a watercourse could be shown on a map with any straight sections. The elders were scanning the banks closely. Eventually they guided us into a spot where, instead of mangroves, there was a section of steep dry bank about three metres high with tall brown grass, spiky plants and trees. This I understood was the rock wall. The boat crept, reversed and slid in close to the bank.

'Crazy Pete', the deckie, sprang ashore with an anchor line to secure to a tree. I slung my camera bag over a shoulder and followed. But the shifting weight of the camera bag altered my balance, and I missed the firm handhold I expected, slipping down almost two metres over the sandy rock surface before I could grasp a plant and my feet could find a hold. From there I took an easier approach to the top of the slope, where I could view the sea only a couple of hundred metres away, while the expedition party of Ross, Michael, Clive and Francis (despite his sore foot) came ashore. On their way they saw a dead crocodile in the water just below where I had slipped.

The elders pointed down westward to a grove of trees with fresh green leaves — 'the oak trees'. Clive moved off at his deceptive walking pace that obliged us to run at times to keep up, and Francis was limping but he managed to keep up with us much of the time. We reached the white beach, turned south and a bit inland among the clumps of trees, and stopped.

'This was the place where those things happened with the Dutch', Clive Yunkaporta announced confidently. But he was still looking for the spot where the wells were dug. The cyclone winds had changed the look of the country, blown things away and covered up the other wells, he explained.

The elders set off again, now on a zigzagging course through the lengthening shadows among the trees, and they stopped to look all around. Last year there had been some trees at one spot, but they were gone now. They continued walking, searching, at one point warning me to watch out for snakes.

Near here, in the elders' traditional version, some Dutch sailors were buried in 1606.

'Here was the well.' Clive pointed to a grey-brown sandy patch. Just here, his 'long, long ago' ancestor Peter had jumped down into the well and the spearing and shooting had occurred. He turned back toward the beach and we followed while Francis, still searching for the site of the other well beyond the trees, limped back toward the boat.

Clive stopped on the dry white beach about seventy-five metres from the edge of the sea. He seemed to have reached a decision and we waited for his verdict. 'Yes. The Dutch bodies; they buried 'em round here.'

I had travelled by air, sea, land and river a total of nearly 5000 kilometres to hear those nine significant words, and for a few moments it seemed an anticlimax.

But standing there in the glare of the late afternoon sun it became easy to imagine the *Duyfken* anchored about nine kilometres out across the flat blue sea, tiny in the distance, and the dramatic scenes that had been described, even if some of the many scattered details seemed inconsistent with each other or with my own understanding from historical documents. I had not been so naive as to expect to see the whitened bones of the sailors from the *Duyfken*'s landing boat party.

We returned to the boat and headed back next morning for Weipa, studying with more understanding the coast and rivers drawn on *Duyfken*'s chart. Ross Gardner and Michael Fordham agreed with me that it was now obvious that only the ship's boat could have navigated into the rivers. They had no doubt the boat crew, the first Europeans, must have made a number of landings on the mainland, if only for water and firewood, and probably sometimes by accident in shallow patches of the rivers.

'Mission accomplished', Ross Gardner declared.

Jack Spear and the Grassbird Man's Story

AT THE WEIPA Aboriginal settlement of Napranam I was presented with a book entitled *This is What Happened: Historical Narratives by Aboriginals*, edited by Luise Hercus and Peter Sutton.

I was surprised to find that I had been virtually following, nearly twenty years later, in the footprints of Dr Peter Sutton, a South Australian anthropologist who had spent years living in the bush to learn the languages so that he could transcribe the story of Dutchmen at Cape Keerweer for his PhD thesis at the University of Queensland. I also found from Sutton's book that there were a number of slightly different versions of the Aboriginal story of 1606.

The Sutton book quoted primarily an Aboriginal elder named Jack Spear Karntin, who was born in about 1905, a few kilometres inland from Cape Keerweer. Much older than the Yunkaporta brothers, Jack Spear Karntin has since died, but his story was recorded in his own Wik-Ngatharra language at Peret Outstation, about twenty-five kilometres north of Cape Keerweer, on 12 December 1976.

In Peter Sutton's translation, Jack Spear Karntin states that there were no white people in his country before 'those Dutchmen' came to Thewena, the proper name for the mouth of the Kirke River. It was there, not Cape Keerweer, where the Dutchmen landed.

The Dutch built boats and a 'house' there, Jack Spear claimed. The Aborigines did not yet understand ... 'the poor things used to kill each other, used to spear each other'. They were still doing this when the Dutch arrived in their country, according to Jack Spear. If the Dutchmen had behaved properly the Aborigines would not have killed them. But they had detained the two wives of Jack Spear's ancestor called the Grassbird man. He identified them: 'Like, they belonged to my mother's younger brother — from my mother's older brother, whose name was Grassbird; he was one of them — well, he (the younger brother, the Grassbird man) was lending his two wives to the Dutchmen. [Sutton noted that the clan whose major totem was Grassbird was now extinct. Their country was centred right on the mouth of the Kirke. The lending of wives to visiting white men in the early period was not rare.]

'Then all of his cousin brothers came and joined together with Pat [the English name that would be given to the Grassbird man much later], and they arrived at the [Dutch] boat, where food was offered to them. The Grassbird man said: "No ... we've come for our women". Some other cousins said the women were being kept in the boat.'

The Dutchmen said: 'Yes, yes, they're here'. The women came up from the boat and the Dutchmen said they could go.

Then one of the Dutchmen invited an Aborigine to use his gun to shoot some black ducks. The other Aborigines said: 'No, he might shoot you, he wants to shoot you', and they wrenched the gun from his hand and hurled it into the sea.

A Dutchman asked where was the gun, and 'they grabbed him by the neck and smashed him down. They clubbed his nose, then the back of his neck. They hit another, stone dead! Another was starting to come out from down below, and they hit him next, striking the back of his neck and nose.

'The whole lot of them that were in the boat got finished off. They threw them into the waters.'

Three others who were up on the bank were killed at once. Next, the Aborigines incinerated the boat and swam across to the eastern side of the Kirke River.

After this, more Dutch boats came in from way out to sea off Cape Keerweer. These Dutchmen, having wrongly blamed the Aborigines on the south side of the river mouth, shot them with guns as they lay sleeping: 'Bang, bang, bang, bang, bang, bang, bang. But they were innocent'.

The survivors of the punitive attack dived into the water, swimming to the east bank and heading for their homeland. They fled into the scrub, but the Dutch guns followed them.

In this story the warriors had killed perhaps nine or twelve Dutchmen at Cape Keerweer, and if people went back there in 1976 they would have found the well the Dutchmen left behind. 'It was they who dug it out — black people never dug it out.'

Peter Sutton's book stated that when John Gillanders, a past superintendent of Aurukun Mission had taken Jack Spear Karntin and the late Victor Wolmby to the place some years before 1976, they had told him: 'Here it is. It was down there that they were killed, those Dutchmen'. The book had a photo of some Aborigines standing beside what they said was the remains of the well.

The story returned to the issue of the women: 'But they kept those two women too long. It was from that reason. So they got wild with them ... But they didn't understand about friendship, not them. They just went "any way" ... The old people didn't understand, for they had no friends'.

Peter Sutton noted that the Aboriginal figure of nine Dutchmen killed appeared to come ultimately from the Samuel Purchas 1625 publication of the journal of English Captain John Saris.

He also referred to his own visit to the well in 1977, when he recorded a much longer version of the Dutchmen story. Told with tremendous animation and attention to detail by Noel Peemuggina in Wik-Ngathana and Clive Yunkaporta in Wik-Mungkana, this long version still awaited transcription and translation. He found: 'The well is indeed of non-Aboriginal origin, since large chunks of shellgrit have been levered out to make a wide hole. Aboriginal wells are not excavations in this sense. I was also shown where Aboriginal men had been employed to cut a path through the mangroves to allow access from boats to the well.'

In 1904 Moravian missionaries established the mission at Aurukun, and since about 1928, following the death by spearing of four Aboriginal men during a formal cremation ceremony battle, the missionaries had been making trips by packhorse and boat to the Cape Keerweer area, to take children back to Aurukun for European education and to introduce a new form of law and order. They also made enormous efforts to keep other Europeans out of the area, did not force everybody into a settlement and did not in general oppose Aboriginal customs or languages. Reverend Bill MacKenzie, the superintendent for more than forty years between the 1920s and 1960s, learnt the local Wik-Mungkana language, underwent the traditional initiation ceremony and told the people: 'Never let anyone take your land away from you'.

In spite of his aggressively disciplinarian approach, MacKenzie was remembered with respect if not love. During his time Cape Keerweer was mainly free of the killing of significant numbers of Aborigines, as happened in most other parts of Australia. There were massacres of Aborigines in the wider area, and the white man's revenge had a long reputation for being excessive, indiscriminate and efficient, as Jack Spear's story revealed.

At a community meeting in Aurukun on 12 December 1975, while deliberating on whether to shut down the mining camp of the Anglo-Dutch oil company, an elder named Billiton got up and said they had defeated the Dutch before at Cape Keerweer, and they could do it again.[60] Peter Sutton observed that the Cape Keerweer people were aware of their special place in the story of Aboriginal contact with Europeans. They frequently told of how they had repulsed the Dutch, and it was a story told with passion.

Jack Spear's version of the story made specific reference to a published account of the seventeenth-century Dutch visit to Cape Keerweer, and in view of the depth of time involved, Dr Sutton said, we must conclude that this work had influenced the way the story is told. The story possibly had other possible foundations, in a remembered clash with white men at Cape Keerweer in the late nineteenth or early twentieth century, since the Aboriginal participants known as Pat, Pumpkin, Bob and the Grassbird man were young or mature men at that time.

'Was Mr Karntin's version of the events at Cape Keerweer historically accurate?', Sutton pondered. 'Without better documentary evidence from the period of the lugger contact on the coast around the turn of the century, this we cannot judge with much precision. Some such evidence may one day be unearthed, and there is documentary evidence for the view that such occurrences were not uncommon at the time.' There was a contemporary report, for example, that detention of Aboriginal women on boats north of Cooktown had led to disputes.

But the basic tenet of the story was undeniable: 'There were indeed cases in Australian history when Aboriginal resistance forced a European retreat'. In the earliest phase, which was dominated by the Dutch, Aboriginal weaponry was only marginally if at all inferior to that of the whites, and their numbers were greatly superior. Taken in conjunction with the dangerous seas, the superficially barren landscape, and the difficulty encountered in finding water along the Gulf coast, the highly effective retaliation of the Aboriginal inhabitants must have been the clinching deterrent to Dutch colonisation of North Queensland.

It was not surprising to Sutton that the 'Dutchmen at Cape Keerweer' story had reached high popularity among Cape Keerweer people in the present situation. 'They are threatened as never before.'

Peter Sutton raised an issue which had been concerning me from the beginning of my own historical research. 'For what European historians would consider to be a reliable account of the initial exploration period, we must turn to documentary sources', he wrote. Then he turned to the observation of the historian Andrew Sharp:

> Now when an objective historian says that such and such a thing happened, or did not happen, he means parenthetically 'according to an objective view of the known evidence'. In that sense, which is the only sense in which history can be said to have any validity, it can be stated quite firmly that the first European discovery of Australia was when the Dutch vessel *Duyfken*, after coming along the south coast of New Guinea from the west, swung over to the west coast of Cape York Peninsula in or about March 1606.[61]

Sutton continued: 'Ironically however, our first such record consists entirely of secondhand hearsay! Having been printed, it does of course enter the academic shrine of "the literature" '.

Secondhand hearsay? Perhaps worse, thirdhand, I calculated, recalling that the English writer Samuel Purchas in 1625 had quoted the journal of English East India Company agent in Java, Captain John Saris, reporting that he had been told on 15 June 1606 by Captain Tingall from a Java junk, that the exploring Dutch ship [*Duyfken*] had returned to Banda after finding the island of New Guinea but had lost nine of her crew 'killed by the Heathens, which are man-eaters ... '.

Heathens? My conclusion by then was that the Aboriginal people of Cape York Peninsula of the seventeenth century would have been deeply religious, their 'particular system of faith and worship' as defined in *The Australian Concise Oxford Dictionary* being a collection of spiritual beliefs that linked them with the land of their ancestors and ordered their lives. As for 'man-eaters', I had found no evidence in the historical records to justify declaring the Wik people of 1606 to be cannibals.

At Aurukun (I was advised that its name came from a missionary's mispronunciation of an Aboriginal term meaning 'place where plenty fresh water sits down') I was given a tape recording called 'Listening to the Aurukun Story', produced in 1991 for the Aurukun Community Corporation and the Aurukun Shire Council. On this tape Topsy Wolmby, Bruce Yunkaporta and Silas Wolmby gave another similar version of the Dutch landing in 1606.

One difference from others I had heard was that when the Aboriginal people of Cape Keerweer first saw the Dutch ship they took it for a big pelican. One of the Aborigines stated at the end: 'This is an old story, but it's new today'.

When the tape recording was made, about 1100 people lived at Aurukun, ninety kilometres south of Weipa along with approximately 100 non-Aboriginal employees and their families. Aurukun Shire Council was established by statute in 1978 as a mainstream local government authority. The council was the trustee of 7500 square kilometres of land within the shire, held under a fifty-year lease. However, enactment of the Aboriginal Land Bill was to result in this land being converted to inalienable freehold title.

The council had all the municipal and planning functions of a Queensland local authority, plus a number of other functions that would not normally be undertaken by a local government, including provision and maintenance of houses it owned, being all of the housing in the Aurukun township, plus social services, community employment and training, and law and order.[62]

'Listening to the Aurukun Story' expressed regret that the recent impact of the introduction of alcohol had contributed to an increase in serious problems in the community. These problems have been compounded by the greater awareness people had of the wider community and their access to it, of involvement in activities and a way of living that was inconsistent with the maintenance of the cultural life and practices. But the outstations offered some hope for the renewal of traditional social, economic and cultural practices. These would ensure that the land and places of spiritual importance were cared for.

BACK AT WEIPA, the expedition members prepared for a journey by road of about seventy kilometres north to Mapoon and the Wenlock River, named in the seventeenth century as the Batavia River, where it was recorded that one Dutch sailor from *Duyfken* had been fatally speared.

But before this journey, I wanted to construct my image of what the Dutch might have seen when they first gazed upon the Australian coast in 1606. The most practical way to achieve this, I decided, would be to fly low over it in a light aircraft.

The *Duyfken* chart showed her first landfall at a river which seemed certainly to be the Pennefather, fifty-two kilometres roughly north-north-west from the Weipa airstrip. The river was named after a Queensland mariner of the 1880s, Captain C. Pennefather.

Ross Gardner and I boarded a high-wing, single-engine Cessna at the airport for a charter flight over the area. I had explained to the pilot that I wanted to examine and photograph the mast-height view the Dutch would have had as *Duyfken* was first approaching the Australian coast. The aircraft's right-hand door was removed so that I could lean out of the side to observe and photograph; a bit windy but quite convenient.

We flew out across the coast and turned to run northward about five kilometres from shore, and there, beyond the serene blue sea dappled with grey shadows of puffy cumulus clouds, was the beckoning sight of the 'River with the Bush' as *Duyfken*'s chart of 1606 had named the Pennefather River. She had sailed in from the north-west and probably anchored about four nautical miles out while the ship's boat was rowed in toward the creamy white strand, several hundred metres of sand with a line of small trees behind it.

Ross Gardner was tapping my left arm and shouted into the wind against the flat roar of the Cessna's engine: 'No mangroves there, mate'. I nodded happily. Speech was difficult, but it was simple for me to direct the pilot with hand signs to swing back and then ease down at lower altitudes for more runs up close to the mouth of the river. I wanted a nearer look at that 'bush' the Dutch had identified when they came ashore here.

At the southern side of the narrow entrance to the river we could see a welcoming yellow sandy lagoon, its slender southern seaward arm linking it with the coast. Looking southward, we found more trees — but not the mangroves that lined the other rivers I had seen on this coast.

We banked to port for the last time, turning back to fly down over the land, a stretch of grey-green bushes and then low trees. From just above, we could see the thin black tree trunks reaching out of white sand revealing an image of a hungry, thirsty land that the Dutch had dismissed. No good to be done here.

Where the coast opened up into Albatross Bay, with Weipa in the eastern distance, Duyfken Point was difficult to pick from three sandy points. I chose the second one, which was holding a little tidal lake and reaching out into the shallows where little wavelets were breaking.

There was said to be a monument at Duyfken Point signifying that the Dutch had sailed past in 1606 but we could not see it. One of my photos would show the end of my boot, resting on the leg of the undercarriage as I peered down to search for the point. Some local people were convinced the Dutch had come ashore here and named it so, but it was Captain Matthew Flinders, RN, who named it in 1802 as a tribute to the Dutch explorers. It seemed most unlikely that the Dutch would have brought the ship's boat inside Janssen Shoal, about three miles out, without a good reason.

The Cessna slipped down through a bumpy south-easterly Weipa wind to a smooth slow landing.

MAPS SHOWED roads from Weipa to Duyfken Point fit for a four-wheel-drive vehicle, but none to Pennefather River. Locals said there was a track leading out to the coast south of the Pennefather and you could probably drive up the beach to it, but we decided to drive first to Mapoon and the Wenlock River and to try the adventurous Pennefather River track on the way back.

The road for Mapoon was fine for about thirty kilometres through thirsty-looking trees and plants to where it ended, meeting a track running up on the western side of the Wenlock River. The track was reasonable until it began to curve north-east towards Cullen Point and took us closer to the river. From time to time the meandering grey sandy surface degenerated to such a soft sinking texture that it could not seriously be called even a bush track. We arrived in the late afternoon at the main administrative office building of the vague sprawling settlement of Mapoon, with its hospital, school and varying types of houses. The reception at this sleepy outpost was friendly; they had exchanged telephone and fax messages with me from Perth and had received letters recommending cooperation with my research. We were advised to talk to an Aboriginal ranger who lived up the road a bit, and meanwhile they would try to find some elders, including several women who were senior voices in the story of old Mapoon.

We set off hopefully with somewhat loose northerly directions toward distant Cullen Point. The winding track led us through dunes, past houses and close to the sandy river bank where a few people were fishing in the dying sunset.

Past a lone house the winding track took a sharp, blind turn and ran down toward the beach and then beside it. The track surface sank alarmingly, bringing us to a noisy stop. We backed and went forward on a slightly different course but this took us even closer to the water and halted us about seven metres from where the tide seemed to be coming in.

We got out and searched for loose dead tree branches to place behind the back wheels to firm up the surface for a reverse escape. These were scarce and difficult to spot in the fading light, but I found what looked like coconut palm fronds under a tangled clump of bushes, and urgently hauled some out to place across the track. The vehicle roared, slipped and then gradually slithered in reverse out of the soft sand trap and managed to reach firmer ground where we could turn back. In the near darkness, utterly exhausted from pushing our bogged vehicle, we headed back for the house we had passed, to ask for directions to where we might camp in our tents for the night.

We were in luck. The occupant stepped out to greet us: 'G'day. I'm Lawrie Booth, the Aboriginal ranger'. We told him who we were, and he wouldn't hear of us going somewhere to camp for the night. We should come in and have a beer and a yarn and some dinner if we hadn't yet eaten, and sleep in the house. It was clear he would be offended if we declined his hospitality. When I went to our car fridge to get some cans he told me not to bother; he had plenty. But I insisted, ours would be really cold.

Lawrie introduced his wife Sharin, daughter Heide, aged nine, and son Brett, eight. All were keenly interested in the conversation which developed, about *Duyfken* and the European discovery of Australia. Lawrie was an ex-soldier and versatile in his skills acquired from various jobs. At this house he had his own power plant and water supply. He was not an elder and said he could not speak as such, but some things were known about this area by all the people. Turning to his own background, he told us he had discovered there was an English aristocrat, a baron or a lord, on one side of his distant ancestry.

As he vaguely understood the story of the Dutch at the Batavia (Wenlock) River, it seemed the Dutch had come here wanting to be friendly but also to grab some blacks and take them away with them, an odd mixture of motives. There was some fighting with spears against the Dutch blunderbusses (muskets). The Dutch had also come in to the Skardon River, about twenty-seven kilometres north of Mapoon, Mr Booth believed. But in the last century the infamous Jardine brothers had come through this area and shot a lot of Aborigines. Many Aboriginal skulls and bones had been found in recent years. Lawrie Booth agreed that it might have been the 1623 Carstensz expedition that visited the Skardon River.

While we were talking I was conscious of something crawling on my arm. I brushed it casually, thinking it must be a mosquito or fly. The feeling continued, and I looked down carefully; it was a flea or tick. I flicked it away on to the concrete floor and tried discreetly to squash it under my boot. A few moments later the crawling sensation was resumed. More of them, definitely. I looked around, but everyone seemed at ease. Michael Fordham was busy cooking dinner for us and our hosts from supplies he brought in from our vehicle. All agreed that our cool Riesling was splendid.

Then I felt the creepy crawling presence in the hair of my head. That was too much. I asked Lawrie if I could have a shower because I felt grimy after scraping around in the sand when we were bogged. 'Sure', he said, and added that I would find hot water and soap and towel there.

While showering I remembered those palm fronds I had dragged out to place across the track near the river edge — they would have been carrying the crawling creatures which had swarmed off and transferred to me. I recalled warnings I had heard about coconut palm branches when I was in Papua New Guinea in 1942. Some Australians had died from what was called scrub typhus, contracted when they touched the palm fronds where the tiny insects lurked. After the shower I had no more problems. Incredibly, I must have washed all the crawling things away down the drain. I rejoined the conversation.

In the morning Lawrie Booth and his family were up early. They had arranged to drive away for a holiday in the bush, down at a place called Stony Crossing, on the southern reaches of the Wenlock River — I remembered a notice on the roadside. There was a nice pool for the kids to play in, but Lawrie said he would walk the banks for signs of crocodiles and watch the water for an hour before he, and then the kids, would swim in it. It was a few miles inland, so I hoped that if any, they would be freshwater crocodiles and not the savage saltwater man-eaters.

'But the house is yours as long as you want to stay', he said. 'Just shut the doors when you leave.' We thanked him and said we would be leaving soon after him to keep up with our schedule.

Before they left, I was taking a photo of the happy children Heide and Brett, when they volunteered to sing a song for me:

> I'm an Aborigine, make no mistake.
> I'm an Aborigine, make no mistake.

Just a couple of lines, but there was a wealth of meaning. It touched my heart. I clapped hands to express my pleasure that they showed pride in their ancestry.

We drove back to the Mapoon community office where we learned that Susan Madua, the senior woman who might have talked to us, had gone off into the bush for a few days. She would stay out hunting turtles, wallabies and wild pigs, and digging for yams, they said. But Malcolm Koitop, a senior member of the community, a friendly, modest man whose ancestors had come from the Torres Strait Islands, took us to the house of his uncle Jack Callope several hundred metres away. We sat on chairs in the shade of a tree outside Mr Callope's house, while the birds kept up an almost constant lively chorus of chattering from tree to tree. Soft-spoken Jack

Callope, whose grandfather had come from the Solomon Islands, was evidently accepted for his Aboriginality, but although he asserted this as a member of this community, he was also reluctant to be taken as a formal representative of the Mapoon people.

Heide and Brett Booth of Mapoon, who sang their song of hope, 'I'm an Aborigine, make no mistake'.

He was aware of stories about the Dutch at the nearby Batavia/Wenlock River — 'That was Dutch guns against my people's spears'.

But he was more concerned with the frightful events of last century, especially the massacres attributed largely to the infamous Jardine brothers, Frank and John, in the 1860s and for several decades following. I had read various accounts of how the Jardines' merciless operations cleared out all the Aborigines from north of the Ducie River — which joins the Wenlock at Cullen Point — except for demoralised survivors gathered about missions at Lockerbie and Somerset around the top of the Cape.[63] Lawrie Booth told me some people would say it was Lachlan Kennedy and not Jardine who shot the blacks at Namaletar River, and that these blacks had been shot by their own people or those of another tribe.

Jack Callope described to me an area not far away, where evidence of the frightful past was revealed fairly recently after salt water flooded over the bank of the Wenlock River and exposed human bones: 'Lots of bullet holes in the bones'. I understood him to mean that these were Aboriginal victims of the Dutch in 1606. But it was what he saw at another place further inland where he had gone mustering on the Namaletar River, running into the Skardon River, north of Mapoon, that had shocked him:

> At that spring I was told you must be very careful not to move those bones, and I said: 'What bones?'. I was told: 'You go and have a look; they went to get a drink of water there because they were thirsty'.
>
> I saw those bones, they been shot in the head, and the bones were in the creek. Three points on that creek, one, two, three, the third point; that's where they got shot. Where they went to the spring for water. I felt sick when I looked down through the spring water and saw all them bones ...

There was a strange silence now outside the rambling old house, and it seemed that even the birds had ceased their chatter to listen closely with us to the sad story of murdered men's whitened bones.

> A church man wanted me to take 'im up there and show 'im. Was all right in the morning, but I was told: 'Don't show nobody, don't take nobody there'. And next day I say to that church man: 'No, won't take 'im there. I got to be very careful, not touch ...'.
>
> On the streams; that's where you find bones. No, not the Dutch did all that killing of my people there. These white men, English [nineteenth-century colonists] they said: 'This our land'.

How many skeletons? 'About fifty.' Jack Callope was becoming increasingly uneasy discussing the subject of death. 'My people don't like to talk about these things. Don't want to say where these bones are. People might come.' Ross Gardner said for us: 'We got respect for those things, we don't want to go there'.

In 1965 the Queensland Government drove all the Aboriginal people out of Mapoon, forcing many of them to go to Thursday

Island, but the people started coming back to Mapoon, we were told. It was not the mission that wanted them out, but a mining company that wanted the area for white sand mining. The mining company had been allowed to work on sand mining at a white

Malcolm Koitop (left) and Jack Callope tell the story at Mapoon.

mission in New South Wales, and 'they thought maybe they would be allowed to do this at Mapoon now, but the Aboriginal people of Mapoon, they say no, so they went to Skardon River'.

Hours later, carrying with us many other stories of Aboriginal history and legend, we drove away with grinding low gears through sinking grey sand stretches, past scattered rambling houses of Mapoon, some looking very makeshift, near the historic Batavia/Wenlock River. We realised from occasional stops to ask directions and brief conversations with the locals that there was a lingering hostility to white visitors, what I sensed as a suspicion that we strangers might be a few more of the whites who would once again try to drive them away from their ancestral lands.

The story of *Duyfken* and the Dutch discoveries had been overtaken on this cape by the more recent bitter legacies of British colonialism, but the music of hope lingered on: 'I'm an Aborigine, make no mistake'.

About forty kilometres down the track to Weipa, we turned right, onto a track heading west for the coast. We had been advised that this led to the beach and that it was possible to drive a four-wheel-drive vehicle the eight kilometres up to the mouth of the Pennefather River. I wanted a close examination of the trees at 'River with the Bush' that was marked on *Duyfken*'s 1606 chart.

The track proved to be reasonable until it ended about 100 metres from the beach, in an area shown on my chart as 'Low sand hills'. The blinding white sand dunes, however, were much higher and softer than we expected, and after a couple of hours of charging unsuccessfully up the side of veritable mountains of burning sand we accepted that it was impossible to plough through to the beach with this vehicle. The way north was similarly blocked.

Standing sweating, defeated, on a ridge overlooking the sea, we saw a distant flurry of beach sand moving down the beach from the direction of the Pennefather River. It was an aging but brave Landrover, its engine and transmission screaming as it battled past us at about twenty kilometres per hour. We waved a greeting from 100 metres, fully realising that the driver could not dare to stop or even slow down in this endless mass of fine flying sand. Then we cooled down with a cautious swim in the warm water of the Gulf (but not past waist depth) and imagined *Duyfken* sailing past, several miles out, 389 years ago, her unimpressed crew eyeing the low shore of 'Nova Guinea'. I wrote a message in the sand: *Duyfken* passed here 1606.

In Weipa next day I suggested to some of the townspeople that they should call this remote northern coastline Australia's 'Coast of Discovery', where the nation's history stood still waiting for us all to read its mysterious messages. I telephoned Isobell Savo and Sandy Suchet at Napranam to thank them for the cooperation I had enjoyed from the Wik people at various places on Cape York Peninsula.

X

HMAS *Flinders* and *Duyfken*

ON OUR RETURN from Weipa to Cairns I was surprised to learn from Ross Gardner that the Royal Australian Navy had a vessel named the *Duyfken*, a survey motorboat that works with the Navy's hydrographic ship HMAS *Flinders*.

After some hasty maritime lateral thinking and discussion about the significance and potential of the name I decided to write to the Navy with a request that it participate with me in a joint survey of the 1606 *Duyfken*'s Australian discovery course. I suggested using HMAS *Flinders* and her SMB *Duyfken* to follow the Willem Jansz chart up the coast of Cape York Peninsula and into Torres Strait, and explained that the proposed expedition was associated with the *Duyfken* Foundation's plans to build an authentic replica at Fremantle.

I sent my letter, dated 1 November 1995, to the appropriate senior officer of the Navy, Rear-Admiral C. J. Oxenbould, OA, at Maritime Headquarters in Sydney.

A fortnight later I was pleased by the positive reply. The Navy had agreed to undertake the *Duyfken* 1606 Rediscovery Expedition in April 1996, to retrace the original course, except for some inshore sections with navigation safety issues in the poorly charted waters. That was fine: I had already covered the shallow sections from Weipa south to Cape Keerweer, and it was the sections north of Weipa and into the Torres Strait area that needed to be done to complete the study.

I contacted the deputy of Hydrographer Operations, Captain Geoff Geraghty, as advised by Rear Admiral Oxenbould, to gratefully accept the Navy's offer, and afterwards telephoned the commanding officer of HMAS *Flinders*, Lieutenant Commander John Maschke. It was soon clear that he was also keenly interested in Australia's maritime history and happy to be involved in a voyage in the wake of the 1606 *Duyfken*.

To Maschke I sent a copy of the *Duyfken* chart along with a suggested course traced on sections of the modern charts, and we arranged to meet at Weipa on Monday, 8 April.

My Australian-Dutch corporate supporter Fugro Survey agreed to support this 1996 rediscovery expedition, and on Monday afternoon, 8 April 1996, Ross Gardner and I arrived at Weipa Airport to be met by John Maschke, a fit-looking six-footer in Navy white shorts and shirt, who drove us to the little wharf and HMAS *Flinders*. She was a beautiful little ship, gleaming white and tidy, waiting at Evans Landing.

HMAS *Flinders* at Weipa ready to begin the survey of the *Duyfken* Australian discovery chart.

Leaving port next day the ship was humming with activity: the rising note of the two big diesel engines and electrical equipment, stand-by calls and orders preparing for departure, reports to the captain from other officers on the bridge, radio messages, and the 'tok-tok' of the echo sounder. The ship moved away smoothly, turned and headed down the Embley River. As we glided past the Weipa Peninsula a group of sailors were lined up on parade near the bow in what was evidently the Navy's custom in leaving port. It was impressive to watch from the bridge.

HMAS *Flinders*, 49.07 metres (161 feet) in overall length, displacement 750 tonnes, and her 10.7-metre (35-foot) survey motor boat SMB *Duyfken* were coastal hydrographic units which collected data used in the production of marine navigation charts for Australia and Papua New Guinea. *Flinders* was built at the Williamstown Naval Dockyard in Victoria and commissioned in 1973, and the SMB *Duyfken* was one of six boats introduced in 1993 to improve the surveying efficiency of the Navy. Together they provided a specialised platform to aid hydrographic personnel in collecting data necessary to ensure the waterways in Australia's area of responsibility were safe for maritime navigation.

They undertook such hydrographic tasks as determining water depth with positional accuracy of less than ten metres, searching for navigational hazards using side scan and hull-mounted sonars, collecting tidal data, accurately fixing the coastline and compiling or correcting sailing directions. *Flinders* also conducted oceanographic observations including ascertaining the nature of the seabed, determining water visibility levels and conducting bathythermographic observations to determine the sound velocity of the water. She had a cruising speed of thirteen knots.

SMB *Duyfken* was a high-speed craft with a maximum speed of twenty-nine knots and a cruising speed of twenty knots. She used the same data logging equipment as *Flinders* but had only one sonar. The boat could operate independently from the ship to conduct soundings inshore or in areas too dangerous for the ship: something like the relationship the 1606 *Duyfken* had with other ships in the Dutch East Indies fleets.

I looked down astern from the bridge at SMB *Duyfken* sitting under her launching gear on the port side of the deck, with a dinghy nearby. The other dinghy with the mooring party came alongside and was shipped while *Flinders* was under way. Our departure from Evans Landing had been delayed by some pieces of floating mangrove or palm branches blocking a seawater intake and we needed to make up some time. The ship produced ten tonnes of fresh water daily for its engines and domestic use. We turned into the dredged South Channel, marked by piles and beacons for about twelve kilometres, but swung out of it after about six kilometres to head more to the west and then north to clear Duyfken Point.

According to *Australia Pilot*, the mariners' guide published by the Hydrographer of the Navy, the Duyfken Point warning light was exhibited from a white metal framework tower, twenty metres in height, close to the southern extremity of the point, and there was a monument between the point and the light structure. But because of the drying reefs and foul ground extending out for about two kilometres from the point, we were too far away for me to spot the tower or the monument.

Captain Matthew Flinders had sailed down this coast in 1802 in HMS *Investigator*, and in his book *A Voyage to Terra Australia*, undertaken for the purpose of completing the discovery of that vast country, he referred to 'DUYFHEN' [sic] Point on 8 November:

> This point is one of the very few remarkable projections to be found on this low coast, but it is not noticed on the Dutch chart [probably that of Tasman in the 1640s]; there is little doubt however, that it was seen in 1606, in the yacht DUYFHEN, the first vessel which discovered any part of [the Gulf of] Carpentaria, and that the remembrance may not be lost, I gave the name of the vessel to the point …

In the afternoon several menacing grey weather fronts reared up in the west of the Gulf, and as each approached the order went out to close up ship. While all watertight doors were shut tight, heavy rain swept the ship and rushed onward to the land, leaving decks running and dripping with fresh water under a hot sun and clear sky.

The sea was calm blue and we were in deeper water about ten kilometres from shore, when the loudspeaker system announced an assembly of the ship's company, other than those on watch. About thirty-five stood together in a wide circle on the deck near the bow, the ship travelling steadily north about eleven knots. The commanding officer addressed the crew on the research purposes of this voyage — with maritime author James Henderson, following the course of *Duyfken* of 1606 as shown on the seventeenth-century chart, from Weipa to Pennefather River and the Wenlock (then Batavia) River, up into Torres Strait.

The executive officer Lieutenant Rod Shepherdson, standing between me and the captain, tapped my shoulder and told me I should respond now. So I named some great navigators who had sailed in these waters: the Dutch captain Willem Jansz of the *Duyfken* early in 1606; the Spanish Luis Baez de Torres who sailed past Cape York about seven months after Jansz had departed; the Dutchman Abel Tasman in the 1640s; the English James Cook in 1770; William Bligh in his longboat after the mutiny on the *Bounty* in 1789; Matthew Flinders in *Investigator* in 1802. And the captain of our HMAS *Flinders* was upholding the Navy's fine exploring tradition.

From my studies for the replica ship we would soon be building in Fremantle, I said, *Duyfken* 1606 had emerged as an exciting historic ship as well as the first European one here. 'This *Duyfken* 1606 Rediscovery Expedition is the first such voyage officially undertaken to survey the seventeenth-century discovery course of the Dutch ship *Duyfken*, using that ship's chart', I said, holding up an enlarged copy of the chart, which fluttered in the light breeze.

'This is itself a historical voyage in which all of us here today on HMAS *Flinders* are involved, one I believe we can speak about with pride to friends and relatives. I feel honoured to be associated with the Navy's great Hydrographic Service in researching Australia's earliest maritime history.'

I briefly explained the name and reasons for 'River with the Bush' on the Dutch chart, said I hoped to get ashore there, and offered to answer questions. The applause was pleasing, and the varied questions confirmed that all were glad to be involved in the expedition.

Afterwards I asked John Maschke if he could join in the visit to Pennefather River. 'I wouldn't miss it', he said, and I learnt that detailed planning was already in progress. One of the dinghies was being prepared for launching on the starboard side. About 4.45 p.m. the ship came to anchor and the crew pointed shoreward to the mouth of the Pennefather River, but I found it difficult to discern under the build-up of clouds there.

On 7 November 1802 Flinders landed at the Pennefather River, and wrote that the small opening at latitude 12°13′S appeared to be 'the Coen River of the Dutch chart'. This undoubtedly was today's Pennefather River at 12°14′S. Captain Flinders gave its longitude as 141°47′E, which placed the mouth 5.6 kilometres up the river, but in his time it was still difficult to fix longitudes accurately. The entrance to the river was too small and too shallow to admit anything larger than boats, Flinders wrote, and commented that soon after his boat had shoved off from the shore a party of sixteen 'Indians', all naked and carrying spears, had appeared and called out to the departing Englishmen.

Ross Gardner and I were handed life jackets, raincoats and helmets, all to be donned before stepping into the boat. With the addition of my camera gear slung over the shoulder my movements were somewhat encumbered. The commanding officer and a couple of sailors joined us and the dinghy was lowered to make for the shore. One of the sailors, Lance Sheward, was armed with a high-powered rifle — another safety measure. He explained that it was a precaution against a crocodile attack, but his instructions were that he was to use it only if necessary.

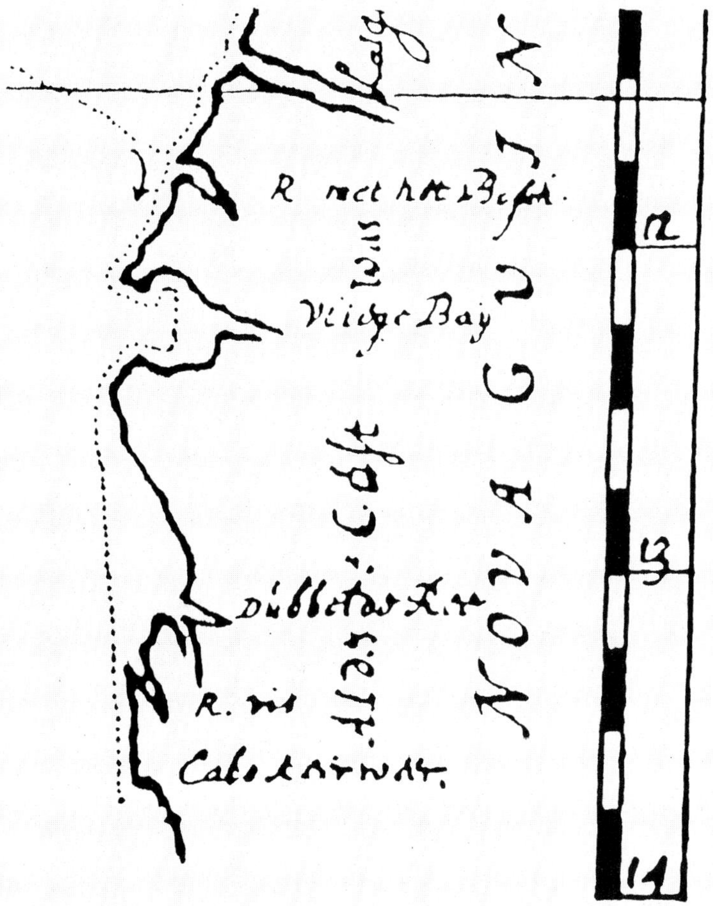

This enlarged section of *Duyfken*'s chart shows where she first approached Cape York Peninsula at Pennefather River about March 1606 and sailed south to Cape Keerweer before returning to Banda.

About half-way to the shore the forty-horsepower outboard spluttered and stopped. After a sailor had checked and cleaned the spark plug the motor still refused to fire. Maschke called the ship with a mobile telephone, spoke to the engineering officer and applied his suggestions for the outboard, but it remained dead. The captain called the ship again and gave instructions for another outboard and a spare to be brought to us.

Time was running out. The light would soon be starting to fade, I expected. We waited: a situation in which a captain might well have shown impatience, perhaps expressed some profanity towards someone to be blamed. But the skipper remained quite calm. Not one angry word.

In the distance the sun glinted gold flashes on the side of the other dinghy as it moved away from the ship and took shape speeding toward us. We quickly transferred to it and left the other dinghy's crew to go back using a spare outboard they had brought.

We steered for the coast on the northern side of the river and soon the dinghy grated in over the coarse pale brown sand that had shone bright yellow when I saw it from the air on my previous visit. I stepped ashore without getting my socks wet and hastened to the river bank to examine the trees, the key to this page of seventeenth-century Australian history. Lance Sheward followed closely in my tracks lest I tread on something savage in the shadows.

The dying sunlight held on long enough for me to confirm certainly that it was not mangroves here on both banks of 'River with the Bush' but a species of casuarina like the sheoaks of Western Australia. Bundles of these would have been first rate for the cook's stove on the *Duyfken*. Some types of Indonesian mangroves were considered suitable, I was told later, but evidently *Duyfken*'s Dutch crew had considered the trees they found to be of significance for their chart.

'Congratulations, skipper', I said to John Maschke, 'you've just rediscovered *Duyfken*'s Australia', and Ross Gardner photographed us shaking hands to celebrate the event.

It was almost dark when we pushed off the dinghy. Unlike Captain Matthew Flinders in 1802 we did not see any 'Indians', as we left here and headed for the last glow of the sunset. The captain called HMS *Flinders* on his mobile telephone to let them know we were returning. Their radar would have picked us up.

The elegant white ship at anchor loomed up before us in the black sea, a beautiful glowing sight — I could understand the skipper being very fond of her. He told me that because the delays had left us with such a short time at the important river mouth he would revise his schedule to allow us another visit in the morning. He would not come but would send one of his officers with us.

In the morning I looked ashore and was surprised to see how clearly the mouth of the Pennefather River was outlined under the rising sun. When we arrived there in the dinghy we cautiously entered the river through the doubtful channel, wary of the shallows.

I had read *Australia Pilot*'s warning that 'the entrance cannot be distinguished until abreast of it', and as usual the advice was correct. We went about 100 metres up the river, speculating on how far the Dutchmen had sailed in their ship's boat. Their chart showed the river running from about seventeen kilometres east, a bit longer that the modern chart showed it, but they would have seen it in the wet season. We wandered about the banks as the Dutch would have done when looking for firewood and water, escorted again by Lance Sheward, but no crocodiles showed.

Ashore at 'River with the Bush' (Pennefather River), the first Dutch landfall, author Henderson congratulates LCDR John Maschke of HMAS *Flinders*: 'You've rediscovered *Duyfken*'s Australia'.

We headed back to the waiting ship and were brought back aboard without delay. The anchor was raised and we resumed the northerly course, thirty-five kilometres up to Cullen Point for the Wenlock River.

Next morning the ship was anchored about twelve kilometres out from the coast. It was time to bring our *Duyfken* into the action. Lieutenant Commander Maschke told Ross Gardner and me that *Flinders* would raise anchor while we were ashore and continue north; we could catch up in the motorboat.

The land at Cullen Point, at the north-west entrance to the rivers, was densely fringed by mangroves. The chart showed a sandbank with a reef at 0.2 metre (at low tide) on our starboard bow to be avoided, and *Australia Pilot* referred to a somewhat intricate channel leading to the river mouth. Our boat had plenty of modern navigation equipment but the crew took it very cautiously, crossing a bank marked as only 0.3 metre, to reach the channel leading into the junction of the eastern Ducie and the southern Wenlock rivers.[64]

A southern course took us into the Wenlock/Batavia River, surrounded by thick mangrove banks. After several miles Ross Gardner and I were looking ashore at a distant sandy beach and scattered buildings.

The survey motor boat *Duyfken* anchored in the Wenlock River. A *Duyfken* crewman was recorded as speared at this area during a landing in 1606, most likely at the sandy beach.

We thought it was along this very beach that, as noted in the journal of the 1623 Carstensz expedition, a sailor from *Duyfken*'s boat had been killed in 1606 in an encounter with the Aborigines. Unlike other parts of the Wenlock River with forbidding stretches of thick and seemingly endless mangroves, the beach appeared suitable for a landing here.

We launched the dinghy from the stern of SMB *Duyfken* — not to go ashore but only to enable me to get some space to photograph the Navy's survey motorboat on the river. In this gloomy environment it was not difficult to visualise the violent incident. Afterwards the Dutch would not have wasted any more time here.

Back at our *Duyfken*, we shipped the dinghy, and the navigator carefully negotiated our return to open sea, heading westward for about nine kilometres and then north, avoiding a shoal marked as 0.6 metre, and soon passing the hazardous Kerr Reef (1.2 metres). HMAS *Flinders* was ahead, cruising at about ten knots. We caught up, slid in beside her aft deck, and enjoyed the smooth, efficient procedure in which the ship's davits plucked SMB *Duyfken* from sea while in transit and swung her into her resting place on the deck. Glancing down while we were airborne,

I was surprised by the afternoon colour of the sea, the deepest emerald green I had ever seen.

The 1606 *Duyfken* chart indistinctly showed two or three more small openings in the coast, the first about nine kilometres north of here presumably being the Skardon River, and then a couple more rivers, perhaps the Jackson and the Macdonald. But the Dutch must have viewed them from the masthead in passing, rather than risk sending the ship's boat in for a more careful survey. We were also keeping our distance, well out from Vrilya Point and turning to 337 degrees (about north-north-west) to avoid the rows of very shallow banks extending out about twenty-five kilometres from Slade Point at the western end of what is called Endeavour Strait, where actually Torres late in 1606 was the first European known to have sailed between Cape York and Prince of Wales Island.

The Dutch chart showed what seemed to be Crab Island a couple of miles out from Slade Point and the word 'Moent' on the land near it, but I could not find that word in my Dutch-English dictionaries. The *Duyfken* chart also showed what is now Woody Wallis Island and Red Wallis Islet and also another island, but this might have been a sandbank exposed at low tide.

There could be no doubt about 'de Hooghe Eylandt' (the High Island). This had to be Prince of Wales Island, and a dotted line on the *Duyfken* chart indicated that the ship's boat had been sent eastward about fifteen kilometres to the island, probably to look for water and firewood. The veritable nightmare of Rothsay Banks at the western exit of Torres Strait would have kept the ship well out.

The *Duyfken* chart marked about a dozen more islands north of here, the bigger Badu and Moa islands easily identified. But in HMAS *Flinders* we followed our north-north-west course to come to anchor for the night about one nautical mile off historic Booby Island, the westernmost island of Torres Strait. An automatic warning light shone from the uninhabited island.

We went ashore at 6.30 a.m. in a dinghy with John Maschke, the executive officer and the buffer (chief bosun's mate), Mark Cooper. The commanding officer, who had last called there in 1994, led us about eight metres down a steep cliff to a deep cave which sailing ships of the last century used as a sort of mail depot and message centre. Sailors had painted their ships' names and dates on the walls, and at least one Second World War naval ship was represented. (Any fresh names would earn Navy disapproval.)

Back on the ship, we weighed anchor and turned back northward to resume what would have been the course of Willem Jansz and *Duyfken*, through the area described on the 1993 chart (Aus 376) as 'Inadequately Surveyed'. The Dutch had drawn the Badu and Moa Islands east of here, recognisable on the *Duyfken* chart, and William Bligh (on his way with the longboat to Timor after his mutiny on HMS *Bounty*) had navigated a precarious shallow course westward around the top of Badu, through what still bears the name Bligh Passage.

Farther north our modern chart showed we were passing west of an area littered with shoals at 1.5, 1.6, 0.6, 3.3 and 0.3 metres. We could imagine the exasperation of *Duyfken*'s Willem Jansz, forced back westward time and again from his attempts to sail north to reach what might be the distant 'Nova Guinea' coast. With a draught of only 2.1 metres he would still have been sending his ship's boat ahead with a sounding lead to check the depths. The boat crew could have climbed over the side and waded in some places.

A few miles north of here our chart showed 'Discoloured Water' and one of the shoals at only 2.1 metres. Also about here, on my copy of *Duyfken*'s chart of 1606, was a dotted outline, about fifty kilometres long and roughly east-west, of what was termed 'Vuyle Bancken', meaning 'dirty shallows, and sandbanks or reefs'.

In the distant east-north-east about sixty kilometres away on our naval chart was Turnagain Island, another reminder perhaps of Cape Keerweer (Turn-about) on Cape York Peninsula, but it was a bit too far east of the Dutch course for *Duyfken* to have reached it in 1606, and was virtually barred by a series of sandbanks, coral pinnacles, and shallows. Probably Carstensz named it so in 1623.

After about thirty-five kilometres on his west-south-west course in retreat from Vuyle Bancken, Willem Jansz would have seen more trouble ahead, a dark strip about four kilometres long on the western horizon: Turu Cay at 9°49.5′S, 141°26′E.

HMAS *Flinders* came to anchor about six kilometres from this narrow strip of coral and sand. From the distance it had appeared as a dark brown line, but closer it became a beckoning yellow strand and near its western end could be seen a ship, sitting crazily almost high and dry and abandoned. SMB *Duyfken* was launched to search and repair a tidal gauge that was reported to have broken from its mooring not far away, while the captain decided to go ashore accompanied by the ship's navigating officer Lieutenant Mike Moore, two sailors, and me, to have a look at the cay and the grounded ship.

I learnt that she was a Taiwan fishing vessel, about 1200 tonnes, and she had been there several years. Apparently she had lost power during a violent north-west monsoon and was driven almost to her full length on to the beach. Frayed fishing nets were still drooping over the port side near the bow — perhaps the crew slid down them to into shallow water during a wild night, and monstrous following seas had swept the ship up onto the land. On the beach on her port side was an anchor.

We waded around the stern in water almost up to my armpits, and John Maschke pointed out some coils of heavy rope tangled tightly around her shaft and propeller. Probably this was the cause of the disaster.

Before returning to the ship I dropped my camera gear and clothes on the burning hot sand and indulged in the luxury of a swim about seventy-five metres out in the clear water. It was magnificent. Back on HMAS *Flinders* I reflected that the swim had been the most remote one of my life, nearly sixty kilometres from the Papuan coast and twice that distance from Cape York.

After SMB *Duyfken* radioed that the tide gauge had been located, HMAS *Flinders* left the SMB with the task of repair and observation and proceeded west-south-west on the course of the 1606 *Duyfken* to where I guessed Captain Willem Jansz and his officers had finally decided to give away the search for a northern exit from this wretched Arafura Sea. From here, they had headed safely westward along nearly 10°S latitude until it was appropriate to turn north-west to find their outward course from False Cape of Nova Guinea.

We figuratively farewelled Willem Jansz and his *Duyfken* several miles south-west of Turu Cay and kept on that course for a remote spot on our chart, Glamis Castle Shoal. HMAS *Flinders* resumed her survey work on this stretch and I had the opportunity to learn something about the work of the Navy's Hydrographic Service.

I read from Hydrographic Service information papers that the RAN Hydrographer was responsible to the Chief of Naval Staff as the national hydrographic surveying and charting authority and for the formation of policy to meet Australian hydrographic needs and international obligations.

The Great Barrier Reef is the largest coral reef in the world, containing some 2900 reefs and stretching more than 1900 kilometres from Central Queensland to the shores of Papua New Guinea. It holds a number of navigational hazards to shipping, including fast-flowing tidal streams within the reef, limited water depths, reduced visibility in the wet season, cyclones and narrow restricted shipping lanes.

The Australian *Navigation Act 1912* requires ships to carry up-to-date charts appropriate to the voyage being undertaken; the 1974 Safety of Life at Sea (SOLAS) Convention of the United Nations requires contracting governments to provide navigational information services to mariners; and the 1982 UN Convention of the Law of the Sea (UNCLOS) requires international cooperation to publish information to safeguard life and property at sea and protect the marine environment.

A list of recent groundings on uncharted hazards in Australian waters shows the need for the activities of the service. Here are just some of them:

- 1971: Oil tanker *Oceanic Grandeur* hit an uncharted rock in Torres Strait. The resulting pollution destroyed twenty pearl farms.
- 1978: General cargo vessel *Transgold*, 11,000 tonnes, grounded on poorly surveyed area near Rosade Spit, Papua New Guinea.
- 1980: Cruise liner *Linbad Explorer*, 28,000 tonnes, hit an uncharted rock in the Antarctic.
- 1981: Western Australian State Ship *Kimberley*, 12,000 tonnes, grounded off Derby outside surveyed area. The ship broke her back on the falling tide.
- 1982: Oil tanker *Ampol Sarel*, 78,000 tonnes, grounded in Prince of Wales Channel probably due to shifting sandbanks, highlighting the need for re-surveys in areas known to be unstable.
- 1984: Bulk Carrier *Iron Curtis*, 70,000 tonnes, drawing 12.2 metres on passage in recommended Great Barrier Reef route near Lizard Island, crossed a 12.2-metre uncharted reef. Fortunately it was high tide.
- 1984: Car carrier *Bright Ace* on passage from Japan to Australia grounded on an uncharted shoal in the vicinity of the Star Reefs, Papua New Guinea.

- 1986: HMAS *Betano* was holed by an uncharted rock while surveying in Napier Broome Bay, Western Australia.
- 1986: HMAS *Brunei* grounded in poorly surveyed water in Endeavour Strait while on survey duty.

The bad luck of HMAS *Brunei* reminded me of the Spanish Captain Luis de Torres whose ship bumped on a shallow patch in that vicinity in the night of 3–4 October 1606 but apparently escaped without significant damage.

Paradoxically, hydrographic vessels are perpetually looking for shoals, but to physically find one with the ship's hull would very likely be a disaster, John Maschke explained to me. Most ships aren't designed today to allow them to touch the bottom. Like the West Australian ship *Kimberley* in 1981, they are likely to suffer broken backs or similar doom. An example of naval black humour was found in the saying that 'every ship has the ability to be a mine sweeper, once'. Similarly every ship could be a crude hydrographic ship once — and could then spend the remainder of its career as a hazard marker.

A number of significant marine pollution incidents, including the *Exxon Valdez* off Alaska and the *Kirki*, off the south coast of Western Australia, considerably raised public and political concern about the risks of accidents resulting in pollution in Australian waters and particularly in the sensitive marine environments of the Great Barrier Reef and Torres Strait regions. To minimise the risk of accidents leading to ship-sourced oil pollution in these regions, the Australian Government gained the support of the International Maritime Organisation in 1990 for declaration of the Great Barrier Reef as a Particularly Sensitive Area.

Several times in the voyage between Turu Cay and Glamis Castle Shoal we saw drifting buoys and lines from long-line fishing boats, obvious night hazards to shipping, and Ross Gardner told me these tuna lines could be more than fifty kilometres in total length with 100,000 hooks, and gill net drift nets over forty kilometres long.

I noted from a report[65] commenting on international infringements of Australian fisheries and the need for official charts for the proper management, policing and necessary prosecutions that:

> International poachers are not sunk on sight and it is necessary to prosecute in a court of law in which the onus of proof is on the prosecutions

and where the accused is entitled to mount a legal defence. In the absence of accurate official charts such prosecutions could be exceedingly difficult.

To give just a couple of examples of the money that can be involved: In 1990 two Japanese long-line tuna boats were found to have been under-reporting their blue fin tuna catch (which is subject to a quota). The first by half and the second by 30%. The master of the first was banned from fishing in Australian waters for two years and ordered to pay costs of $4635. The second boat and its catch and equipment were forfeited to the Commonwealth and its owner and skipper fined $10,000 each.

In the first case the catch was valued at $3.5 million, and in the second $6.3 million.

On the bridge one morning the executive officer relayed a message inviting me to join the commanding officer at the crow's nest. I slung my camera bag over my left shoulder and clambered up the ladder, about thirteen metres but it seemed like nearly twenty metres from the water. From the platform it was an interesting perspective of the ship with its twin white prop washes running straight back to the horizon. The captain told me he thought I might be interested in the view of the ship as it turned into a loop. He telephoned instructions to the executive officer on the bridge and soon our heading began to swing gently to port, from our south-west course to south, south-east, east, and north-east.

The stern view through my camera lens was dramatic. On one side, the boiling foam raged in a glittering arc across the dark blue swells, and on the other side the wash was a parallel trail of light green water sweeping in to join the calm inner circle. The commanding officer told me he liked spending time up here; sometimes it was a good lookout for spotting reefs. About 100 metres behind us, a curious big seabird with a wingspan of around 2.5 metres seemed to be gliding along in pursuit on a stream of warm air from our ship. A strange little white object floating on the dark swells about seventy-five metres astern caught my eye. I recognised my cap that must have blown away while I was photographing the ship's wake. 'Don't worry', the captain said, 'we'll find you another'.

At Glamis Castle Shoal our ship was to check on a recent report of a hazard. We were searching for the position of the shipwreck but Maschke explained: 'It's abstract; you won't see anything outside. The echo sounder will search for it'. I learned later that nothing was found.

Back on deck later I was presented with a smart navy blue cap with the name HMAS *Flinders*.

We had now completed the investigative task of filling in, as far as we could, the blank spaces in the significant history of the Willem Jansz and *Duyfken* discovery and charting of Australia.

Ross Gardner and I were summoned to an unexpected meeting of the ship's company where John Maschke, on behalf of the crew, expressed pleasure and pride in having undertaken the joint historical research expedition with us in the wake of the 1606 *Duyfken*. He presented me with a handsome solid metal copy of the official crest of HMAS *Flinders*, bearing the hydrographic motto 'INVESTIGATE'.

HMAS *Flinders* headed for Thursday Island in Torres Strait to drop off a sick sailor and his medical escort, and Ross Gardner and I also disembarked there, to be welcomed at the wharf by the genial resident naval officer at Thursday Island, George Graham. The ship returned to Turu Cay to pick up SMB *Duyfken* and resume the hydrographic survey, and we flew back to Cairns next day.

Later I learnt that the aging HMAS *Moresby* and *Flinders* were to be paid off in a couple of years and replaced by two new hydrographic ships using a modern, highly sophisticated hydrographic survey system. The total contract for the new ships and their initial support was said to be worth about $200 million, and both were expected to be in service by 1999.

XI

Willems River

IN WESTERN AUSTRALIA another expedition was still to be organised to complete my historical research into the *Duyfken* story: where was Willems River, where the former skipper of the *Duyfken* came ashore from another in 1618?

On 6 October 1618 Willem Jansz wrote to the Amsterdam Chamber of his company, reporting his visit to that part of the Western Australian coast then known as Eendrachtsland, discovered by Dirk Hartog who sailed past in 1616 in the ship *Eendracht*. Jansz was the upper-merchant on *Mauritius* which first came within sight of land near Point Cloates, south of Exmouth, having followed the 1611 Hendrik Brouwer southern route across the Indian Ocean from South Africa.

Jansz wrote after his arrival at Bantam:

> This serves only to inform you that we passed the Cape of Good Hope on 8 June [1618] in the ship *Mauritius* with strong westerly winds, so that we did not deem it advisable to call at any land. We sailed 1000 mijlen to eastward [about 7000 kilometres] ... in latitude 38 degrees south, though we would have wished to go still further East.
>
> On 31 July we discovered an island and landed on same, where we found the marks of human footsteps. On the west side it extends NNE and SSW; it measures 15 mijlen in length, and its northern extremity is in 22 degrees S. Lat. It bears SSE and NNW from the south point of Sunda at 240 mijlen distance [1780 kilometres]; from there we arrived with the help of God at Bantam on the 22nd of August.
>
> Done on board the ship Wapen Van Amsterdam, 6 October 1618. Your Worships' Obedt. Servant WILLEM JANSZ.[66]

Following this report, a 1618 engraved map revised to 1628 by cartographer Hessel Gerritsz showed the Dutch discoveries from 1616 to 1622 on about 1550 kilometres of the western coast of Australia, with the Sunda Strait and part of Java and Sumatra above. The map showed, a little north of 23°S, what could indicate North West Cape and Exmouth Gulf, and below it the name 'Willems Rivier'. The actual river course was not shown.

The name, more likely a celebration of the name of assassinated Dutch leader Prince Willem of Orange than Willem Jansz, seems

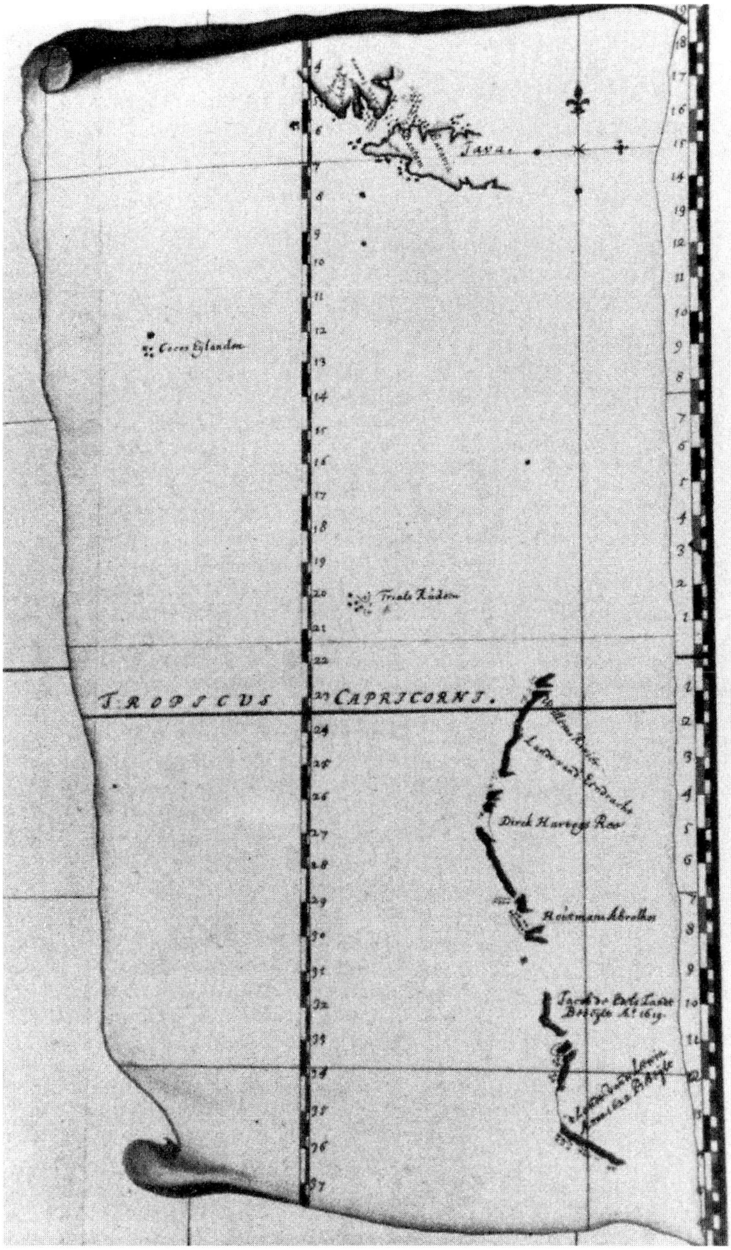

Map of Dutch discoveries along the western coast of Australia, 1616 to 1622, showing Willems River. (From the Van der Hem Atlas, Osterreichische Nationalbibliothek, Vienna, courtesy ARA, The Hague.)

to have vanished from charts after the British settlement in Western Australia.

The latitude of North West Cape is close to 22°47′S, but an error of twenty-three kilometres in measuring latitude in the stormy month of July time off the peninsula of North West Cape was certainly not unreasonable.

In July 1997 Fugro Survey agreed to fund my expedition in search of the lost Willems River, and its marine manager Grey Roughan was happy to accompany me.

Historians had searched in vain for an island '15 mijlen in length' at the position given by Willem Jansz, and the popular choice for the river was the Ashburton — across Exmouth Gulf near the town of Onslow.

Yardie Creek on the modern chart emerged as the most likely position of Willems River, and at Exmouth, Doug Bathgate of Gascoyne Development Commission, with whom I had been discussing the subject for several years, had reached the same conclusion. But this required to be tested, viewed from the appropriate distance at sea, along with the other favoured positions. We first had to examine the issue of an 'island' and then that of the river.

For a start, the seventeenth-century Dutch measurement of one 'mijl' was about four nautical miles, so if there was an island of fifteen mijlen to be seen it had to be at least sixty nautical miles, or 110 kilometres long. Willem Jansz had used the same measurement for his 240 mijlen distance (960 nautical miles or 1780 kilometres) in his north-north-west to south-south-east bearing from his 'island' to Sunda Strait, and come reasonably close: it is about 1940 kilometres from Exmouth to the Sunda Strait.

The Gerritsz map of 1628 did not show any island '15 mijlen in length', we noted, but it did show a break in the coast which would correspond to the shape below Point Cloates, despite some error in the latitude shown there on the Dutch chart.

There was no such island, of course. But the significant clue was the distance north from Point Cloates to Exmouth Gulf — was that peninsula of about 115 kilometres Willem Jansz' island?

Certainly the modern chart showed that the land broke away to the east below Point Cloates, but the question was whether it would have disappeared from sight of people on a sailing ship a safe distance from shore here, suggesting the southern end of an island.

The Exmouth sea rescue group had agreed to its ten-metre boat voyaging down to Point Cloates to participate in expedition, but there was a lot of territory to inspect to complete the necessary research. The practical answer was, as it had been for me at Cape York Peninsula in 1995, to charter a light aircraft for a quick reconnaissance of the relevant area.

On Tuesday, 8 July we took off about midday from Learmonth, the airport south-east of the township of Exmouth, and headed first for the mouth of the Ashburton River about 100 kilometres to the north-west.

It soon became obvious that this could be eliminated as 'Willems River'. There were so many islands and breaking reefs

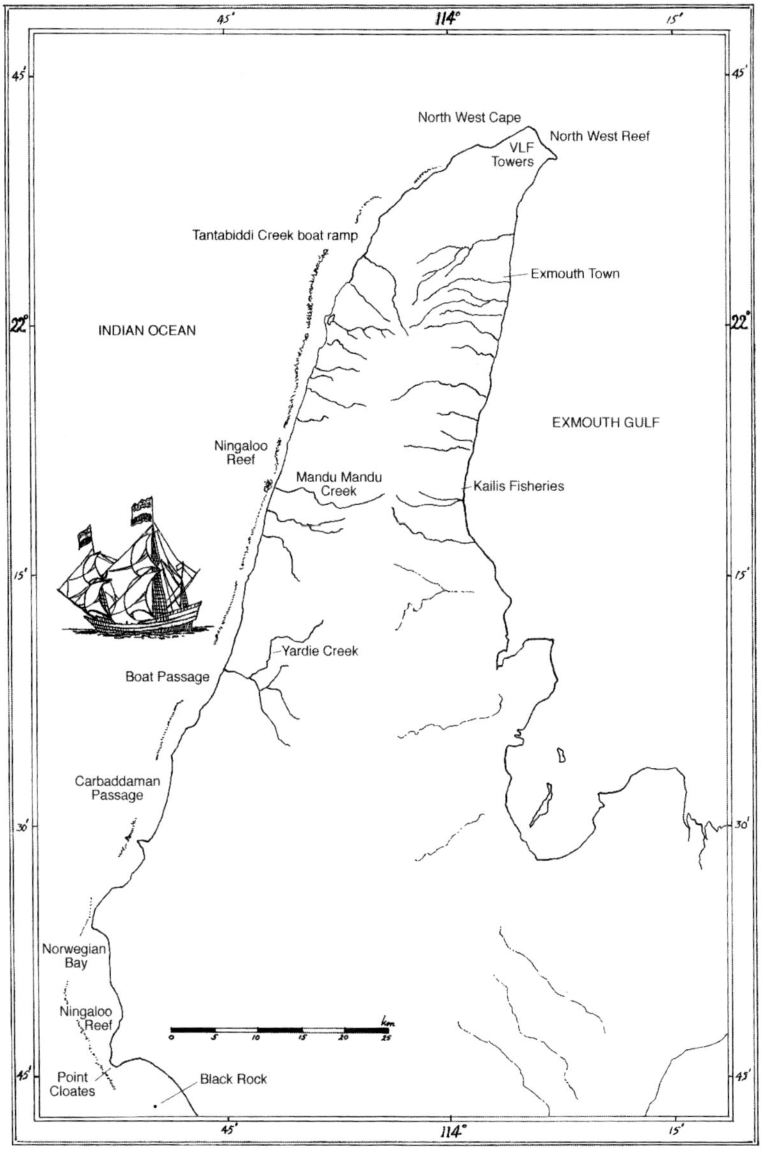

The search for Willems River from the 1618 visit of Willem Jansz, formerly of *Duyfken*, 1606. Author Henderson and Fugro Survey's Grey Roughan located it at Yardie Creek.

cutting off the area from open sea that Willem Jansz' ship would not have risked sailing close enough to send the boat into shore. In any case, the ship would already have turned away from the menace of the cape's North West Reef and from the Western

Australian coast to head for Sunda Strait about 1800 kilometres to north-north-west.

The next relevant observation was of the North and South Muiron Islands, a total of about thirteen kilometres long, running south-south-west to north-north-east in the mouth of the Gulf and thirty kilometres from the mainland coast. Nobody could have mistaken this narrow, shallow passage between two islands in open sea for any river. Below the Muiron Islands, North West Cape and Exmouth Gulf clearly showed me the appropriate shape for the top of Jansz' 'island'.

From here we flew southward down the western coast about 110 kilometres to observe Point Cloates. Seen from about 120 metres altitude and eight kilometres seaward — outside the heavily breaking Ningaloo Reef and the spuming Black Rock — the land did indeed turn away to the east, visually breaking up and fading. Returning northward a little lower, as we passed the point the southern coastline disappeared completely, as expected.

By now we had seen, from something like a ship's masthead view, both ends of what must have been the western side of Willem Jansz' 'island', north-north-west down to south-south-east, fifteen Dutch mijlen long, in July 1618.

Now the issue was the river. It had to be within an appropriate break in the fearsome Ningaloo Reef that rages and roars three kilometres offshore from south of Point Cloates up to North West Cape: this reef makes a vivid, unforgettable spectacle of ocean violence and beauty. The first break in the reef was Carbaddaman Passage, but this would have been too risky for a sailing ship's boat, and there was no river anyway.

Up another eight kilometres or so was another break in the reef. My 1985 marine chart (Aus 745) showed 'Boat Passage' about two kilometres wide, with depths which would have permitted the landing boat of the ship *Mauritius* to negotiate her way into the calm inshore water. Looking down as our plane circled, I reasoned that after sailing about one mile toward shore, Jansz would have encountered shallows that forced him northward before he could conveniently come to the sandy beach. Not far away to the east was higher land, with hills in the distance. He walked north, and there it was: a river he had not seen from the ocean because there was a sandbar across the mouth.

Flying farther north we observed narrower breaks in Ningaloo Reef where the ship's boat could possibly have come ashore, but there was no river or creek. At about 22°9.5′S the Navy's Hydrographic Service chart showed another small break in the reef, and the tiny Mandu Mandu Creek above it, but this did not seem a likely prospect.

From this aerial observation, Yardie Creek appeared as the most likely location of Willems River, but the conclusion needed to be checked from seaward.

The Exmouth sea rescue organisation had arranged to combine its participation in our Willems River expedition with a long-range exercise to test the effectiveness of its navigational systems and radio communications. These links included a recently installed repeater station on Tower Zero of the Harold Holt VLF communications station and another at Point Cloates. Fugro Survey made a generous donation to the voluntary rescue group and also supplied some extra equipment for the boat.

Initially, we planned to depart from Exmouth Sea Rescue Headquarters at Bundegi anchorage early on Wednesday morning and spend two days on the survey, but persistent strong south to south-easterly winds coming up the western coast forced a delay in the schedule. Sea rescue commander John Sandison took the boat on a truck about twenty-seven kilometres around and down to a mooring at Tantabiddi Creek so that we could go aboard there on Wednesday evening for a departure in the early hours of Thursday morning.

The weather showed little abatement, with winds south and south-easterly up to about thirty knots. Mr Sandison, who had a longstanding back injury and was unable to join the expedition, had arranged for a competent member of the sea rescue team, Brian Snook, to skipper the *Rescue Endeavour*, with another member, Mark Damion, to assist. Others aboard with me were Grey Roughan of Fugro Survey and his two teenaged sons Gordon and Stuart, both experienced in small boat handling. Gordon took the dinghy ashore about midnight to pick up Mark Damion.

Grey Roughan had brought with him some Fugro global positioning system (GPS) equipment for the exercise, in addition to the GPS, radio, radio direction-finder and echo sounder installed on the boat.

During the night, while at the mooring, we were contacted by sea rescue radio about a boat (about seventeen metres) which had grounded on the side of a reef several kilometres north of us, and soon afterwards were called by the boat skipper. Contact was maintained, and several hours later the tide rose sufficiently for the boat to float and motor into the Tantabiddi anchorage near us.

We departed about 3 a.m. in inky darkness, Brian Snook skillfully navigating the Ningaloo Reef boat passage on instruments. We had decided that, although it was likely to be very rough and uncomfortable against the high wind and sea, we should make straight for Point Cloates and call in at Yardie Creek on the return.

Grey Roughan of Fugro Survey, who took part in the Henderson Willems River expedition.

Dawn was breaking as we sailed south of Point Cloates in increasing southerly headwinds and rising seas. It was necessary to keep well out to sea in the rough conditions to avoid the hazard of Black Rock, seen breaking heavily about five kilometres south-south-west of Point Cloates and about nine kilometres out from the low sandy area below the Cape.

We turned, with great relief from the buffeting, and observed the disappearance of the coast east of here below the higher land backing Point Cloates — presumably Entrance Hill of sixty-seven metres and Sextant Hill, eighty-seven metres.

On 31 July 1618, the skipper would have been keeping the square-rigged *Mauritius* safely well out from Ningaloo Reef, and now he must have swung even further seaward to avoid the string of warning breakers on his starboard bow. Here, without doubt for us, was the southern end of the 'island' recorded by Willem Jansz. Now we were following him northward, nearly four centuries later, toward the long-lost Willems River.

Still outside the Ningaloo Reef but much more comfortably with the wind and savage seas contesting behind us, we could imagine the Dutch sailors looking shoreward, shuddering at the sight of these long lines of white water breaking continuously and regularly exploding into silvery towers of spray hovering over some of the higher parts of the reef.

There was an opening of about four kilometres width in the main Ningaloo Reef at 22°26′S, but the risks of a group of individual reefs and shoals would have eliminated this as a boat passage for Jansz, and here beside Carbaddaman Passage there was no river anyway.

Willem Jansz, formerly of *Duyfken*, landed here in 1618 and followed footprints up the beach to what was to be named Willems River — shown to be today's Yardie Creek.

Running at about five knots before this southerly wind that had made our voyage down to Point Cloates uncomfortable, *Mauritius* probably came into sight of the next break in the Ningaloo Reef in the easing conditions of early afternoon.

From the safety of the 100-metre bank nearly nine kilometres out, some higher land could be seen to tempt Willem Jansz to sail ashore with the ship's boat and to set foot on this stretch of land discovered by the *Eendracht* in 1616. Twelve years before this time, on the *Duyfken*, he had seen his boat crew go ashore at a number of rivers of 'Nova Guinea', but now as the upper-merchant he could rely on the skipper Lenaert Jacobszoon to worry about the safety of the ship. The ship's landing boat was launched …

Our sea rescue boat *Rescue Endeavour* came in to anchor about seventy-five metres from shore. Here, at 22°19.5′S by the Navy

Hydrographic Service chart, the almost landlocked Yardie Creek suddenly beckoned brilliant blue from behind the yellow sandy beach, with a shallow river overflow a couple of metres wide spilling over the sandbar. We waded ashore. Jansz' boat passage would have brought him in south of Yardie Creek, and we knew he found some human footprints which soon led to what would be called Willems River. We followed the Dutchmen's course and found the same, except that they were not Aboriginal tracks.

Thus Yardie Creek was not disqualified just because it could not be viewed from the ocean, as some people had argued; the assertion was irrelevant. North of here we found no other safe-looking break in the formidable Ningaloo Reef where Jansz might have sailed through in the ship's boat, to walk up the beach and find a river.

The conclusion we brought back from this 1997 expedition was that there was no serious alternative to Yardie Creek as the site of Willems River where Willem Jansz came ashore in 1618. We hope the original name of the river will be restored before 2006, for the 400th year celebration of Willem Jansz' discovery of Australia.

En Passant

IT HAS BEEN a unique couple of years, watching some stacks of timber being expertly transformed into a graceful vessel inspiring unqualified admiration from all who come to watch.

There are some enduring images. For instance, the first bulky lumps of imported oak sailing down from a crane to the shipyard area borrowed from the Western Australian Maritime Museum, and our Foundation chairman Michael Kailis swinging an adze to make the traditional first cut. Obviously he means business. We can be confident that the replica of Australia's very first historic ship will be built here at the ideal place. He'll make sure it happens.

Another positive image: a crowd of supporters packed into the shipyard, watching as the Netherlands' Prince Willem-Alexander ceremonially lays the keel — just a few pieces of timber, but it signals to the nation that the work has begun. Suddenly, hundreds of gleaming orange balloons soar above us, reaching out for the bright blue Fremantle sky. What a nice surprise. Nobody here will forget this historic day. The heroic sixteenth-century Prince Willem of Orange, whose life and death I researched, lives on.

During that ceremony the generous Indonesian Julius Tahija, who with his Melbourne-born wife Jean donated the cost of the timber for the vessel, speaks as a statesman: 'Today's ceremony is symbolic of thinking of the past and planning for the future'. The audience listens silently to this soft spoken Second World War soldier who fought for Australia: 'In a world of conflict, let us not dwell on matters of conflict, but specifically focus our wisdom on matters of mutual interest between our three countries — Australia, the Netherlands and Indonesia'.

Some months later, with wood smoke hanging over the corner of Cliff Street, the planking level on the hull has risen and she is starting to look a little like a ship, the honey-coloured oak here and there scarred black from the fire-bending.

An event worth waiting for was the lowering of a huge piece of timber, the keelson, into the skeleton of the ship. Bill Leonard and Nick Truelove move about anxiously to control its placing in the bottom. Crowds of schoolchildren are trooping in to look at history being built; nobody can tell them now that our early Australian history was not exciting. The thudding of a shipwright's sledgehammer, slamming some of the thousands of trunnels into the hull, fastening it all together.

In conversation with Bill Leonard aboard the little ship under construction, you know you're at the centre of a great twentieth-century Australian national achievement. He listens patiently, respectfully, to your questions, and his answers are illustrated with lively movements of face, hands, arms, arresting and good-humoured body language. You sense that our early Australian history so long neglected by the nation is being transformed, partly by his hands, into a vibrant story that people want to understand and embrace as their own. The master shipwright and his team are acknowledged as world leaders in their field, and he talks with great pride about his colleagues' skill and hard work and of their dedication to the historic project. 'It's a loyal team', he tells me, and I know he's dead right about it. Looking around at the elegant shape of the *Duyfken* you can be convinced that it took a man with the soul of a poet to create her.

But my most magical moment comes one day in July 1998, standing on the platform near the waist admiring the magnificent artistic lines curving back from the bows and coming down the lower sheer toward me. The bow is smoothly outlined by the forward part of the rail, and you can still see below the deck beams to the ribs reaching up from the bottom of the hull.

Now I notice just ahead of me, several shipwrights carefully manoeuvring a heavy seven-metre-long piece of oak above the ship's side — the next section of that rail they call the rahout. It is curved and sloped in several ways and I get the merest brief glimpse of a number of holes or slots in the under surface before the shipwrights lower the forward end of the rail toward its position. At my end Nick Truelove is concentrating with muscle and mind, easing back and moving the rail up slightly, outward and then back. Bill Leonard is hovering nearby; even he is still for this rare, long moment. Time stands still for the compelling question.

Then down ... yes: the rail locks instantly into position. With all those critical angles and shapes to meet, it's a perfect fit, something vaguely symbolic in it. I can breathe again, and think about the wonderful skills recreated from the distant past.

Soon (about Australia Day 1999) the Little Dove will proudly exhibit the craft skills of yesterday and today to a huge crowd, when we'll all celebrate the historic launching. It will be just 400 years, we believe, after the *Duyfken* was launched on the other side of the world. What a privilege to have been part of all of this!

So my voyage of discovery of events, from nearly four centuries ago until late 1998, ends as this book prepares to go to press and we plan to launch our *Duyfken*.

Soon we are to pass on to another voyage of discovery.

Readers may watch Australia's Little Dove take to the Indian Ocean and prove that she can sail like her distant ancestor — eventually through the waters of which I have written, and to the shallow Cape York 'Coast of Discovery' where Willem Jansz concluded 'no good was to be done'.

I hope we have 'done some good' here by building our history and its earliest ship for all Australians, and that fair winds will carry our *Duyfken* into the next century with an enduring message of national and international tolerance and goodwill, perhaps even a few 'lasten' of generosity and reconciliation.

APPENDIX 1

The 1608 Booklet — 'The Savage Dove'

IN 1608 a booklet — or perhaps a pamphlet that could be folded, judging by the way it was printed — dealing with the Harmensz expedition was published by the well-known Amsterdammer Cornelis Claesz, and it painted a more vivid picture of the early seventeenth-century sailing ship environment, with a frank account of the ferocious 1601–02 Battle of Bantam.

The works of Claesz included the famous *Itinerario* of J. H. van Linschoten, 1595–96. The 1608 booklet was called the *True Representation of the Miraculous and Memorable Conflict.*

An edited version follows, with some explanations, translated by Adriaan de Jong of the *Duyfken* 1606 Replica Project.

IT IS sufficiently well-known goodwilling Reader, concerning the staple, or trading places of spices such as pepper, cinnamon, cloves, nutmegs, mace, ginger and so on, more of all medicinal herbs, exquisite minerals, as diamonds, rubies, turquoises, emeralds, sapphires and others as well as pearls, with the many uncountable other fine artefacts, which can only be found in the East Indies, and are brought from there to here.

Within the last 100 years, this same staple used to be primarily in the famous, mighty and beautiful city of Venice where the trade took place and from there transported all over Europe, distributed and used in such a manner that the aforementioned city with her mighty republic was in great prosperity, renowned everywhere and held in great esteem because of this trade.

It is true that these spices, herbs and other peculiarities to come to Venice by sea from Alexandria, they first have to be brought into Alexandria by land from India against incredible perils, trouble and expenses because of the very long route but also because of the strong convoys that need to accompany the caravans, to withstand the attacks of Arabian robbers …

Some years ago the stout-hearted actions of the undaunted sea-dogs in contempt of the terrifying perils and terrible storms of the raging winds and sea; with the incredible multitude of wants, sufferings and adversities which they themselves and their companions would encounter and could expect during such far-reaching sea journeys, have opened up new ways, unbeknown to our elders, along which the Indies can be reached by ship.

Namely by sailing all the way down Africa until sighting Cape de bona Esperanza, the name given by her first discoverers, for sake of their steady faith to accomplish their intended voyage to the East Indies from there on, and so on from this cape between the other side of Africa and the island of Madagascar, proceeding along many islands ... the Red Sea, lucky Arabia, the Persian Gulf until Calicut or Goa [the two latter being on the western coast of the Indian mainland].

This way the aforementioned spices and drugs and other rare artifacts and precious goods have been brought from there, ending up in Europe with unequivocally fewer costs and burden and the trade in these goods steadily declined in Venice. In Portugal, in Lisbona it increased to the extent that one can say it was conveyed from one city to another, yes, that it is now completely and almost solely implemented in Lisbona.

Hence all other places and countries trading in such goods have to adjust their markets according to the prices dictated in Lisbona, because the Kingdom of Portugal and in particular the City of Lisbona have since some years ago used the peaceful presence of this trade as if being the sole owners.

This is because they have been the first discoverers and finders of the aforementioned route around the Cape de bona Esperanza, and also because her King has counted out a certain sum of pennies to the Emperor Charles in order to completely buy out the Spanish not to try this route.

Thus have the aforementioned City of Lisbona with the people of the Kingdom gained and prospered so much that they are now mighty and rich with an abundance of wares.

Seeing and realising what incredible benefit was afforded to their Kingdom and her citizenry by this trade and Navigation, they have also sought means and ways to secure more and more the aforementioned seaborne trade, to which end these people have by force of men and weapons overpowered many places, cities, provinces and complete Kingdoms in the East Indies and brought under their rule, which they sustain with permanent settlements and wholly enforce and control by Forts and Castles.

But as well as that, they have made such Treaties with those other Kings, Potentates and Princes in the Indies whom they were unable to overpower, that it seemed (according to the word from there) that nobody was allowed or could be trading there but these people alone.

Despite this, not many years ago some ships of the United Provinces of the Netherlands, following the well-known route of the Portuguese far beyond the often mentioned Cape de bona Esperanza, also got to the East Indies where their experience was different from what the Portuguese fame (described above) posed and implied.

Yes indeed, they found that the Portuguese were not doing too well in appropriating and securing all places and regions where the foresaid spices and other requisites are available, so that the people arriving there would have to return home without cargo.

Because there are many Kings and Islands trading in the foresaid spices and numerous other exquisite and rare items, who were either archenemies of the Portuguese or were of such position and attitude that at the request of the Portuguese they have banned everybody from free trading in her region or pledged sole trade to them ...

The Portuguese, being aware of this, and since they were not amused by the arrival of Dutch ships, have against incredible costs raised a mighty armada of ships, in order on the one hand, with or without gratitude, to oblige or force the Kings and also the free regions and islands where the spices are harvested, not to trade with anyone from elsewhere but with them, or otherwise market their crops of pepper, cloves, nutmeg and mace.

On the other hand, also serving to await the arrival of ships from Holland, to forbid this country to re-provision and get water, and eventually to confront and attack these [Dutch] ships, so that nothing would be heard from them again in Europe.

Then, because the aforesaid armada of ships had taken up its commission since the year 1599, and this confrontation and Romanistic battle at sea did not happen until the year 1601: so we will tell you in short what has occupied this ship-armada the two years in between, because for reasons of novelty and strangeness it is worthwhile telling ...

Claesz now explained that the King and Viceroy of Calicut, the Portuguese from Goa and all of the East Indies nations were adversaries and archenemies of each other on land and at sea, but this king had in his naval service a pirate captain named Cunal who was so successful in his constant pursuit and plunder of Portuguese ships that in few short years he became famous and wealthy.

Cunal had become so powerful that his followers nominated him king, and he set up a stronghold, which he named Fort Cunal, on the Malabar coast (from Goa down to Cape Comorin at the foot of India). There he laid off the yoke of the King of Calicut and carried on his piracy operations with impressive success for nearly fifty years, always triumphing over the Portuguese.

But the Portuguese Viceroy of Goa managed to reach a settlement with the King of Calicut, who reluctantly agreed to declare war against the disobedient Cunal. The armada then set sail with 1500 men under command of the Portuguese Lieutenant-General Lowijs de Gamma to the fort of Cunal.

On the night of 12 March 1599 the general landed all his troops on the shore of a river also named Cunal, and sent his Captain Lowijs de Sylua ahead to commence the assault with 300 Portuguese troops. They crossed the river by means of a series of boat-bridges, expecting to be joined ashore by 50,000 Indian soldiers.

Cornelis Claesz continued:

It was a heated battle and the resistance strong, the heavy guns and shot of muskets and hand-guns cracking vehemently from both sides, such that amongst others Captain de Sylua, whose boldness had given the Portuguese heart, was hit by musket fire in the head. No sooner was his death proclaimed than those remaining went retreating to the boat-bridges.

But these had been abandoned by their guards and were floating loose in the middle of the river, of no benefit to those who had fled.

So the first test of this armada, in 1599, was a failure. 'Portuguese inflated arrogance' was blamed because the advice of the King of Calicut, who participated in the battle, had been ignored.

About 20 March 1601 the Portuguese assault on Fort Cunal was resumed under the command of General Andre de Furtado de Mendoza. The worst of the heat had passed, so to take sole honours they did not want Indian troops to support them. The Cunal fort was captured without much effort. As for the pirate leader Cunal:

They captured him in person and he was put in irons on the galleys. Thus this Cunal rose from a bad man to a king and finally became a galley slave.

Having completed this action from the year 1600, the armada was kept busy in the last part of the year 1601 reinforcing itself with people and two new galleys.

It was planned that the armada should be at Bantam, Java, by August 1601 for its main task of throwing out the Dutch usurpers, but gales and thunderstorms thwarted this until 24 December.

They arrived at Bantam, miraculously ordained by the Lord on precisely that same day the three ships from Holland with the two yachts arrived [in the southern entrance of nearby Sunda Strait between Java and Sumatra] ...

Claesz points out now that Dutch ships usually stowed their guns below deck when they passed the Cape of Good Hope. Normally there would be no need for their use on the long ocean stretch. Probably the Spanish ships also stowed their guns after passing the Cape.

They [the fleet of Admiral Wolphert Harmensz] would have fallen into the hands of the Portuguese if it had not been for an unexpected messenger sent to them by the Almighty, a matter worth noticing, who informed them about everything. It was a Chinese who came out in his prao alongside the ships when these were still out of sight from the Portuguese armada, warning them about the arrival of the mighty fleet from Goa, Cochin and Malacca gathered before Bantam, consisting of thirty sail amongst which were eight galleons and twenty-two fustas [small Portuguese rowing galleys] and galleys [oars and sails] all together well armed.

The smallest of the Portuguese galleons were said to be much bigger than the Dutch flagship *Gelderland* and the *Zeelandia.*

He [the Chinese skipper] said also that the Portuguese intended to lay siege to the town of Bantam on water and land to achieve that nobody could be trading without them.

Claesz stated that three Dutch ships and the two yachts had sailed from Texel on 23 April 1601, accompanied by another seven ships, a total of twelve vessels. But only five had arrived in the Sunda Strait when the Chinese had disclosed all this to them, 'in the belief that by his forewarning the clash with such a mighty armada of ships could be avoided'.

But the Almighty Lord encouraged their hearts such that they resolved otherwise. As soon as the admiral of the fleet, Wolphert Hermensz, understood the information from the Chinese [actually via *Duyfken*] he ordered the ships to drop anchors.

A conference of the Broad Council of senior officers of the fleet then decided unanimously that 'trusting God's help they would stout-heartedly attack the aforementioned armada of ships to dispel them from there and relieve the town of Bantam'.

Following this resolution … all that was necessary for the coming battle made ready, have they on the 25 December, two hours before dawn at a fire signal, all together weighed anchors.

About sunrise they came within sight of the Portuguese armada, which had two galleys on watch immediately under the west end of the island Poulo Penzano, and because they did not in the least suspect such awakening, were under attack by these [Dutch] ships before they could set sail or others could come to relieve them, striking them such that they would have been captured had it not been that our Admiral, who was to fire a shot with one of the stern cannon [of the *Gelderland*] had an accident (the cannon burst) to their great [Portuguese] fortune.

The crew of the fusta, in fear of again being attacked, rowed to windward. Our Vice-Admiral Hans Brouwer [on *Zeelandia*], not aware of this, boldly continued the offensive plan, and breached with shots both sides of the Portuguese ship from Malacca. A certain perplexity was noticed in the Portuguese fleet, which retreated to settle under the aforesaid island Poulo Penzano [Pandjang]. Ours, having understood the accident of the Admiral, also anchored near a certain island.

The 26 December 1601 was of such bad weather that neither side could do anything offensively, but it allowed the Dutch the time to complete all repairs.

The 27 December they again went under sail toward the Portuguese fleet, vigorously firing at one another. Near the island to the far-east of Bantam were two of her largest galleys, which had been on guard. Seeing the Dutch ships, these set sails attempting to escape. Then, with the advantage of the 'slaghboech' [the windward position] the ships *Utrecht* and *Wachter* each came to board one of the galleys. Many Portuguese from the galley clambered over *Utrecht* to save their lives, but because so many of them might overwhelm her, the yacht [*Duyfken*] came to relieve her, so most of them jumped overboard.

The captain of this galley was called Francisco de Souza, son of the Contador Major of Lijcxbona Iuan de Tenes, and with him were 23 Portuguese and also blacks or Indians up to 60. Of these, the captain and some Portuguese were taken on board.

The rest were mostly beaten to death.

Regarding the other galley, boarded and taken by the yacht *Wachter*, her captain, an old man who had faithfully served the King at least 23 years in this region was called Andre Rodriguez Palbota. He did not want to give himself up and be taken prisoner. So he was run through with a pike.

They took an iron gun and three iron bassen [small guns, possibly swivel] out of it and having imprisoned the survivors they set it on fire although it was laden with rice and other provisions.

And so particulars of this armada were obtained bit by bit from some captured on board the aforementioned galleys. It is not outside this scope to tell its story ...

Author Cornelis Claesz then gave details of the armada:

First there were five galleons from Goa, on which were the Admiral, Andre de Furtado de Mendoza, and Vice-Admirals Thoma de Zuza da Reucha, Francisco da Silva de Menezeo, Antonio de Zuza Falead and Don Lopes Dalnyda. From Cochin (India) was one carrack. From Malacca came another large galleon with eighteen fustas, all which of which made it to Seylao [Ceylon] but did not come further, except the one we took with the foresaid Captain Francisco de Zuza Toues.

From Minat [Manay] and Seylao came two fustas and two galliots on which [was captain] Andre Rodriguez Palbota, beaten to death by us ...

From Malacca came two galleons and also another two junks, [one] captained by Francisco de Xaris and also another seven 'bantinnen' [Malay galleys]. They also said that on the whole of their armada were 800 men that were Portuguese soldiers, except the sailors that were all blacks.

Said also that another four ships and a yacht lay in Acheen [Banda Aceh, Sumatra] …

Returning to the account of the battle, Claesz wrote that on 28 December 1601 the wind blew so hard that no sails could be carried, and both sides were forced to lay-to. On the next two days they all remained stationary in the shelter of the Java coast 'a little below the wild bight of Bantam'. The Portuguese armada lay to windward of the Dutch, but despite this did not dare to approach or charge as the Dutch had done when there had been favourable winds and weather several days before.

None the less, to do something they had set two of her [Portuguese] galleys on fire, thinking this would force the Dutch to shift, but before these came near the Dutch ships they had already burnt down. They also stripped two of their fusta, and taking the crews on board the ships, wrecked them with gunfire and ran them aground against the shore.

On December 31 it seemed action was to be resumed. The five Dutch vessels weighed anchor and set sails with intent to charge the enemy despite the fact that the Portuguese were to windward. The Portuguese hoisted their sails in response, and the Dutch took in their sails to wait for the enemy's approach — they did not want to be seen as retreating — and dropped anchor. Then the wind dropped. The Portuguese also halted. In the opinion of Claesz the Portuguese did not dare to board or attack the Dutch ships.

Next morning, the first day of 1602, the Dutch challenged the armada again:

Admiral Harmensz weighed his anchor. Sheeting in the sails he set his course straight for the Portuguese armada, and in his wake followed the yacht [*Duyfken*] with the other three in close pursuit. From the opposite the armada was on a collision course with them such that a clash seemed imminent, but they again bore away, afraid of the Dutch guns. Their [Portuguese] Admiral had the bloodflag [of battle] opening out in the wind, but to no avail: his people could not be urged on any more.

The Battle of Bantam was over and the Dutch fleet sailed closer to shore next day to a remarkable reception. They had suffered little damage and lost only one man with several wounded, according to Claesz.

Having won great honour throughout the East Indies, they stayed here until 12 January while the crew were refreshed and the damage to their ships was repaired.

Although there was plenty of cargo available at Bantam, Admiral Harmensz was anxious to proceed on plan with the voyage to Maluku. He arrived at Ternate on 17 February 1602 to a friendly welcome and promises of trade with the king …

The Almighty Lord cannot be praised and thanked enough for the triumphal victory told above, of which the fruits shall be revealed more and more in time to come, having in no small way diminished the arrogance, reputation and airs and graces of the Portuguese and exalted the bravery of the East Indiamen sailing from these lands to those regions.

Claesz trumpeted on to a pious conclusion:

Because, as the well-intentioned reader can read above in Cunal's attack, the Portuguese reckoned their 300 were on a par with 50,000 Indians, what will the Indians call for now that five merchantmen put to rout such a mighty armada of ships?

To the Almighty Lord alone, Who stoutens the hearts of His servants to wreak great deeds, to act boldly and to be victorious in the end: Shall be the glory, praise and honour unto eternity, Amen.

AMSTELREDAM. By Cornelis Claesz. Book-seller on Water in Scripted books, Anno 1608.

Appendix 2

Tom Mutch, MLA, Pioneer of *Duyfken* Studies

The work of Thomas Davies Mutch (1885–1958) entitled *The First Discovery of Australia*, was published by the *Royal Australian Historical Society Journal*, Vol. XXVIII, Part V, and also privately published in Sydney in 1942.

His biography, written by Martha Rutledge for the *Australian Dictionary of Biography*, reveals not only a noted historian but also a journalist, politician, defender of human rights and a close friend of the beloved Australian writer, Henry Lawson.

As a member of Sydney's Mitchell Library committee from 1924 to 1958 Mutch secured the original seventeenth-century copy of the *Duyfken* chart for the Mitchell Library from Dr F. C. Wieder, who published his *Monumenta Cartographica* in 1933. Mutch traced the ownership of the *Duyfken* chart back to a wealthy solicitor of Amsterdam, Laurens van der Hem (1621–1678) who collected maps and charts as a hobby. Van der Hem's collection included *The Secret Atlas of the Dutch East India Company*, with copies of originals no longer in that company's archives. Among these were the charts of *Duyfken* and also *Arnhem*, which accompanied *Pera* to Australia in 1623 in the Carstensz expedition.

The atlas was publicly auctioned at The Hague in November 1730, with a printed catalogue describing it sheet by sheet, and was purchased by Prince Eugene of Savoy, whose heiress sold his entire library to the Hofbibliothek in Vienna. It was while examining this collection in 1921 and 1931 that Wieder found the *Duyfken* and *Arnhem* charts of *The Secret Atlas of the Dutch East India Company* and through him and Mutch they came to the Mitchell Library.

In 1942 Thomas Mutch urged Australians to be grateful to Dr Wieder for the discovery of the *Duyfken* chart in Vienna because, as Wieder had said: 'Before 1606 no map was drawn on which a real portion of Australia was put down from observations. All prior to that was from conjecture'. But it seems to have been Mutch himself who initiated the historical acquisition.

Mutch had arrived in Sydney with his parents from London in 1887. After his mother died in 1899 he left school and home and worked four years in outback shearing sheds where, as an enthusiastic reader, he absorbed socialist literature and the 'bush' tradition. When he returned to Sydney he joined the staff of the *Australian Worker* newspaper and met Henry Lawson. Mutch and Lawson became very good friends, and Mutch tried to reform the poet from his alcoholism.

In 1909 Mutch organised the Australian Writers and Artists Union which merged with the Australian Journalists Association in 1913. He became president of the State branch and vice-president of the federal council of the association in 1915–16. In 1917 when Mutch was elected as Legislative Assembly member for Botany, Henry Lawson told the electors Tom Mutch 'is the straightest mate I ever had'.

As Minister for Education in 1921–22 he tried to get more Australian content into the schools' curricula, and abolished high school fees. He also created the Child Welfare Department. Mutch was Minister for Education in the first ministry of the Labor Premier J. T. Lang, but was excluded from the Lang Cabinet in 1925 after a number of disputes with the Premier.

In 1933, after publishing *The Early Life of Henry Lawson*, Mutch became deeply interested in the voyages of the early Dutch navigators and got the help of the Netherlands Consul-General to procure and translate documents. In 1942 he presented his paper to the Royal Australian Historical Society on 'The First Discovery of Australia, with an Account of the Voyage of the *Duyfken* and the Career of Captain Willem Jansz'. Later studies associated with the *Duyfken* 1606 Replica Project confirmed his remarkably competent, accurate research.

Appendix 3

Details of the Ship

Keel laid: 11 January 1997 by Crown Prince Willem-Alexander of the Netherlands.

Replica: Dutch 'jacht' circa 1600.

Crew: twenty, ranks — skipper, uppersteersman, understeersman, supercargo and seamen.
Cargo capacity: 25 'lasten' (50 tonnes).
Displacement: 125 tonnes (approx).
Length between stem and sternpost: 20m.
Length overall: 24m.
Beam: 5.6m.
Draught: 2.4m.

Construction

Structural timbers and hull: European oak (*Quercus robur*) from Latvia.
Deck, spars: maritime or pinaster pine (*Pinus pinaster*) from Western Australia.

Details of the rigging

Rig: three masts and bowsprit.
Square lower and topsails on fore and mainmasts.
Lateen on mizzen mast.
Rigging: hemp.
Sails: hemp/flax.
Spar dimensions: all solid piece *Pinus pinaster.*
Mainmast flagpole height from deck: 20m.
Mainmast: 15m.
Main topmast (including flagpole): 8.5m.
Mainyard: 12m.
Maintopsail yard: 6m.
Foremast flagpole height from deck: 17m.
Foremast: 12m.
Foretopmast: 7.25m.
Foreyard: 10m.

Foretopsail yard: 5m.
Mizzen mast: 12m.
Mizzen yard: 12m.
Bowsprit: 12m.
Spritsail yard: 5m.

Sail areas

Mainsail: 90sqm.
Maintopsail: 50sqm.
Foresail: 60sqm.
Foretopsail: 32sqm.
Spritsail: 15sqm.
Mizzen: 35sqm.

Appendix 4

The Shipbuilding Team as at July 1998

Staff shipwrights

- Bill Leonard, Master Shipwright
- Don Cockrell
- Nick Truelove
- Tim Preston
- Igor Bjorksten, Master Rigger
- Danny McDermott
- Gabriele Zugaro (trunnelling, timber preparation and trade experience)
- Tom Kulin (faring hull)

Staff apprentices

- Dane Whittaker
- Karim Kouniali
- Ben Taylor
- John McLernon

Staff trade assistants

- Steven Edwards (scaffolder)
- Michael Barnett (fireman)

Contract shipwrights

- Rod Darling
- Gerard Russel

Volunteers

- Ray Miller (precision work and equipment maintenance)
- Rupert Weller (supervisor of work experience, staging platforms, rails and steps, forest knees)
- Jim Lucas (trunnelling)
- Peter Worsley (trade assisting and trunnelling)
- Ron Snook

Administration

- Noel Robins, Chief Executive Officer
- Peggy Rogers, Office Coordinator
- Con Smit, Chairman of Friends

Research

Nick Burningham
Adriaan de Jong

Ship shop

Margaret Rapanaro, Manager
Nina Pope, Assistant Manager
Ria Sjerp
Fiona Chamberlain

Guides

George Trotter, Manager
Jenny Gibbs, Chief Guide

Broomers (volunteers)

Jim Mottram
Don Edwards
Alan Horridge
Don Stanley
Enrico Gaglia
Paul van Hooft
Val Phillips
John Harcourt-Smith
Peter Jellis

The replica project's Chief Executive Officer Noel Robins recalled that the Western Australian Minister for the Arts (and the Museum), Peter Foss, had been concerned by the stop-start problems of the previous *Endeavour* replica project, so permanent employment was secured for Bill Leonard and his first shipwright Don Cockrell, who had been displaced from the State Department of Marine and Harbours and was due to be retrained as a filing clerk. Mr Cockrell was seconded to the *Duyfken* Replica Project, and he will continue to be a valuable worker in the historic ship replica industry and the planned Fremantle Maritime Precinct.

Notes

1 J. H. Parry, *The Age of Reconnaissance*, Weidenfeld & Nicolson Ltd, London, 1963, pp. 193, 194.

2 Parry, pp. 308, 309.

3 Parry, pp. 62, 63.

4 Leonard Y. Andaya, *The World of Maluku, Eastern Indonesia in the Early Modern Period,* University of Hawaii Press, 1993, pp. 3, 114–116.

5 Parry, pp. 204, 205.

6 Andaya, p. 39.

7 Kal Muller, *Indonesian Spice Islands, Maluku*, Periplus Editions, Singapore, 1993, pp. 30, 31.

8 Andaya, pp. 137–143.

9 Information above relating to the Dutch first and second fleets is from *De Eerste Schipvaart Naar Oost-Indie*, Vol. II (Linschoten Vereenigde, 1925); Isaac Commelin, *Beginning and Progress of the United Netherlands East India Company*, 1646; and J. K. de Jonge, *De Opkomst van het Nederlandsch Gezag in Oost Indie, 1862.* Translated by Adriaan de Jong of the *Duyfken* 1606 Replica Project.

10 From the Linschoten Society, No. LIV, in English, with footnotes in Dutch, also a version in the German language and the (Dutch) Linschoten Book LXX, Part One, Appendices, p. 182, referring to the 1594 *Duyfken*.

11 On the questions of what happened to the *Overijssel-Duyfken* after she returned from her third voyage to the Indies, and why was she not reported if lost, Adriaan de Jong thought it probable she was sold off, perhaps renamed, and disappeared from official records as an unimportant small ship.

12 As ARA Catalogue Numbers 135 and 136. The 105 pages of handwritten text of the first journal were divided into two parts: (1) A day-by-day account of the courses sailed, with directions and weather conditions, as well as other observations and comments; (2) The minutes of the meetings of the Broad Council (of senior officers of the fleet), also including an agreement between the Dutch and Bandanese and a copy of a letter by upper-merchant Claes Gaef which had been left with the people of the town of Ayvelilue on the island of Ceram in the Indies.

13 J. K. de Jonge, *The Rise of the Dutch Authority in the East Indies*, Vol. 2, Martinus Nijhoff, The Hague, and Frederik Muller, Amsterdam, 1865.

14 Cornelis Claesz, 'Book-seller on Water in Scripted books, Anno 1608'. The booklet was entitled *True Representation of the Miraculous and Memorable Conflict, Molucken bound ships and yachts from Amstelredam ... Between Five Galleons and twenty-two Galleys and Fusta* ... Cornelis Claesz is understood to have been the publisher of the famous Linschoten volumes of 1594 which encouraged the early Dutch expeditions to the East Indies. See Appendix 1.

15 J. E. Heeres on p. 3 of *The Part Borne by the Dutch in the Discovery of Australia 1606–1765* (Luzac, London, 1899), wrote that if any reliance could be placed on this report, it proved that in 1602 the Portuguese were acquainted with the south-west coast of New Guinea. 'But considering the fact that the Dutch were

utterly unacquainted with New Guinea, is quite possible that on this point they misunderstood the inhabitants of the parts of Ceram visited by them.'

16 The November 1996 issue of *Fortean Times*, a London monthly magazine, quoted a July 1996 Antara Indonesian newsagency report of a previously unknown tribe of tall, white-skinned people living in a small settlement near the upper reaches of the Wirigar River in the mainly unexplored Bird's Head Peninsula of Western Irian Jaya. According to Antara, a schoolteacher named Cosmos Serang in Bintumi region said tribesmen would disappear into the jungle when their tame parrots warned them of the approach of strangers. *Fortean Times* said most Irianese were dark and related racially to the people of adjoining Papua New Guinea.

17 E. S. De Klerck, *History of the Netherlands East Indies*, Vol. 1, W. L. & J. Brusse NV, Rotterdam, 1938, p. 203.

18 Peter van Dam, *Beschrijvinge van de Oost-Indische Compagnie*, F. W. Stapel, Martinus Nyhoff, 's Gravenhage, 1927, Book 1, Vol. 1, p. 24.

19 J. R. Bruijn, F. S. Gaastra and I. Schoffer (eds) *Dutch-Asiatic Shipping in the 17th and 18th Centuries*, Vol. II, Martinus Nijhoff, The Hague, 1979, p. 166.

20 From the 'Journal of income and expenses, mainly concerning equipping, 1602–08'; 31 October 1603, Item 406; (ARA-VOC: 7142).

21 Bert Westra, 'Geld en Geweld (Money and Might)', in Robert Parthesius (ed.) *Batavia, Reconstruction of an East Indiaman*, Batavia Cahier 5, pp. 43–56.

22 See Bruijn, Gaastra and Schoffer, p. 19.

23 See Phillip Playford, *Carpet of Silver: The Wreck of the* Zuytdorp, University of Western Australia Press, 1996.

24 See also *The Voyage of Sir Henry Middleton to the Moluccas 1604–06*, Hakluyt Society 1943, reprint Kraus Reprint Limited, Nedeln/Liechtenstein, 1967.

25 P. A. Leupe, *De Reizen der Nederlanders naar het Zuidland of Nieuw Holland ...*, Amsterdam, 1868, p. 7. Quoted by T. D. Mutch, 'The First Discovery of Australia ...', *Journal of the Royal Australian Historical Society*, Vol. XXVIII, Part 5, p. 19.

26 In 1899 the Royal Dutch Geographical Society published *The Part Borne by the Dutch in the Discovery of Australia 1606–1765*, by J. E. Heeres, Professor at the Dutch Colonial Institute, Delft. Heeres dealt with 'The Voyage of the ship *Duyfken* under command of Willem Jansz(oon) and Jan Lodewijkszoon Roossengijn to New Guinea: Discovery of the East-coast of the present Gulf of Carpentaria'. Heeres explained in footnotes that Captain Saris' date of *Duyfken*'s departure from Bantam was from the 'Old Style' Calendar and should therefore be adjusted by ten days forward to 28 November 1605, and the date of the visit to Bantam of the Klingman from Banda, from 15 to 25 June 1606.

27 Heeres, in a footnote, p. 47, of *The Part Borne by the Dutch* ... wrote that references in the journals of the ships *Arnhem* and *Pera* in the 1623 Carstensz expedition to Cape York Peninsula 'put it beyond doubt that they had on board a chart of the voyage of the ship *Duyfken*'.

28 From R. A. Skelton, *Explorers' Maps*, Routledge & Kegan Paul, London, 1958.

29 The Kei and Aru Islands were already shown on the 1595 Linschoten chart of the East Indies, probably from Portuguese information. Similarly, 'Os Papuas' was shown there as an island running east to west about 560 kilometres and roughly twice the size of nearby Cenam. But only a strip of southern coast was drawn on the *Duyfken* chart as Os Papuas (not to be confused

with the eastern half of New Guinea, which would eventually be called Papua New Guinea).

30 T. D. Mutch commented on p. 31 of 'The First Discovery of Australia ...': 'It is from the Dutch word "bosch" that we derive the word "bush" as applied to country away from the towns' [!].

31 See David R. Moore, *Islanders and Aborigines at Cape York, an ethnographic reconstruction based on the 1848–50 HMS* Rattlesnake *journals of the ship's official artist Oswald W. Brierly*, published by Australian Institute of Aboriginal Studies, Canberra, 1979. Moore, a curator of Anthropology at the Australian Museum, Canberra, quoted (p. 118) descriptions of various canoes, some being built near Evans Bay, Cape York, with outriggers. A Torres Strait Island canoe here had also a mat sail. *Rattlesnake*'s official artist O. W. Brierly told the story of a Scots girl Barbara Thompson, aged seventeen, who was shipwrecked, rescued and adopted by Torres Strait Islanders in 1844, and lived nearly five years on Prince of Wales Island before being found and taken home by the *Rattlesnake* in 1849.

32 See pp. 42, 43.

33 Nick Burningham, nautical architect of the *Duyfken* replica project, put forward a 'speculative reconstruction' of *Duyfken*'s historic voyage, based on his own experiences of voyages between Indonesia and Northern Australia in traditional vessels without an engine. He thought it fairly certain the ship would have suffered no major losses of crew until they visited Tiuri on the return voyage, after which the style of navigation changed dramatically.

34 Brett Hilder, *The Voyage of Torres*, University of Queensland Press, 1980, p. 12.

35 Andaya, pp. 141–143.

36 See Heeres, p. 6.

37 Hilder, p. 127. From studies of the voyages of both *Duyfken* and Torres' ships, I concluded that *Duyfken* might well have crossed the path of Torres' ship in the night during the five months, January to May 1607, when the two Spanish vessels along with the Portuguese from the island of Ternate were fighting against the Dutch, before Torres sailed north on the final section to the Philippines.

38 De Klerck, p. 209.

39 From the Linschoten Society, Book LXX, Part One, Appendices. (Third voyage equipped by the VOC to the East Indies under command of Admiral Paulus van Caerden) pp. 180–181. Translated by Adriaan de Jong.

40 Camel was a piece of wide-muzzled, heavy artillery.

41 Linschoten Society ... p. 182.

42 de Jonge, Vol. 3, p. 253.

43 Marit van Huystee: 'On the yacht *Duyfken* (1601)', Report No. 105 of 1995 to Maritime Archaeology Department, Western Australian Maritime Museum

44 From Linschoten Vereenigde: Book LXX, Part One ... Paulus van Caerden in 1606, p. 180.

45 Andaya, pp. 153–4.

46 See 'Willem Jansz van Amsterdam, admiraal', from P. A. Leupe, *Contributions to the Linguistics, Geography and Anthropology of the Netherlands Indies*, 1872, pp. 298–336.

47 In 1625 the Dutch West India Company built a fort on Manhattan Island, occupied by American Indians, and founded New Amsterdam there. Later known as the New Netherlands, this was seized by the English in 1664.

In 1667 the States-General (Dutch Government) renounced its claims to the New Netherlands and the city was later renamed as New York.

48 See Appendix 2: 'Tom Mutch, MLA, Pioneer of *Duyfken* Studies'.

49 All departing and arrival dates are from J. R. Bruijn, F. S. Gaastra and I. Schoffer, *Dutch-Asiatic Shipping in the 17th and 18th Centuries*, Vols II and III, Martinus Nijhoff, The Hague, 1979.

50 See Phillip Playford, *Voyage of Discovery to Terra Australis by Willem de Vlamingh 1696–97*, Western Australian Museum, 1998, and *Carpet of Silver: The Wreck of the* Zuytdorp, University of Western Australia Press, 1996.

51 The sixteenth- and seventeenth-century Dutch documents refer to the 'jacht' type of vessel, and the Routledge and Kegan Paul Dutch dictionary definition of that term is 'hunt, chase, shooting pursuit'. In the Nederlandsch Historisch Scheepvaart Museum (the Historic Maritime Museum), Amsterdam, the English-language copy of the Descriptive Catalogue refers to a 'pleasure-yacht' exhibit and explains that the international word 'yacht' originated in the Netherlands, that it was derived from the verb 'jagen' meaning to hunt, shoot or chase, to race ... run into expense[!]. Thus a yacht or 'jacht' was a swift vessel. In 1661 the City of Amsterdam presented a small Netherlands yacht to King Charles II of England and thus introduced yachting to that country.

52 From 'Ships', by Robert N. Rose, in the pictorial *Lure of the Sea*, by W. R. McAskill, Halifax, Canada, 1951.

53 Sir Francis Drake is reported to have used these methods in 1588 to get his battle fleet out to sea from Plymouth to pursue the Spanish Armada.

54 Oxford University Press, Melbourne, 1938.

55 *Onward Boy Soldiers*, University of WA Press, 1992, told the story of the Battle of Milne Bay, in Eastern Papua New Guinea, in which I served as a signals runner, and of my postwar search for a Japanese Zero fighter pilot who was shot down in the battle and captured. This research led to New Guinea, Japan and back to Australia to the former POW camp at Cowra, NSW — and the pilot's grave.

56 On 10 November 1802 Captain Matthew Flinders of HMS *Investigator* was near the latitude of the 1606 *Duyfken*'s Cape Keerweer, and he placed it almost correctly at 13°57′S. The 1992 chart of the RAN Hydrographic Service shows the cape as a small black dot at 13°55.5′. Flinders commented in his journal: 'I could see nothing of a cape here, but the southern extremity of the land, seen from the mast head, projects a little; and from respect to antiquity, the Dutch name is there preserved'. HMS *Investigator* would have been safely out at least fourteen kilometres from shore, beyond the seventeen-metre bank, too far for Flinders to spot the mouth of the Kirke River by the cape. My copy of the *Duyfken* chart of Willem Jansz indicated that he also failed to notice the obscure entrance.

57 I understood that the ship *Duyfken* had only one boat — to assist with anchoring, getting water and firewood, etc. But there was a possibility that could not be dismissed, according to Adriaan de Jong, that she was accompanied by another, smaller boat.

58 Peter Pumpkin was apparently the name bestowed in relatively recent times during the influence of white missionaries.

59 In 1940, with anxiety about Japanese invasion, the Australian Army had an outpost at the mouth of the Archer River, near Aurukun; later a radar station too. In *The Torres Strait* (University of Queensland Press, 1989) at p. 39 John

Singe referred to the abundant food sources that were available for the dense coastal population of Cape York Peninsula that Captain James Cook observed in June 1770. Shellfish were of great importance, and at Weipa great shell mounds nine metres high marked traditional camping grounds used over eight centuries.

60 This appears to have been the story reported in the ABC Radio program, Bill Bunbury's 'Talking History', which led to my 1995 Cape York Peninsula expedition.

61 Andrew Sharp, *The Discovery of Australia*, The Clarendon Press, Oxford, 1963, pp. 15, 16.

62 Aurukun Community Incorporated (ACI), established under the Companies Code, had a board of Aboriginal directors representing each clan group with traditional land. It managed the post office and airline agency and operated a range of commercial activities including the store, laundromat, takeaway outlet, a cattle enterprise, and an airline service. Most importantly, however, ACI was the resources agency for outstation communities in the area, and it directed all profits from its commercial activities to outstation development.

63 See John Singe, *The Torres Strait*, p. 55. Singe quoted (p. 49) a missionary named Lawes of the London Missionary Society who wrote in 1883: 'Nowhere in the world have Aborigines been so basely and cruelly treated as in Queensland — the half has never been told'.

64 Cullen Point is at 11°57′S on the RAN chart. Captain Flinders wrote in his journal of 5 November 1802 that: 'The latitude of this opening is 11°55′ and agrees nearly with that of Batavia River in the old Dutch chart'. In *Australia Pilot*, 1972 (p. 27) I read a report from 1887 of 'a magnificent harbour (named Port Musgrave), into which the Batavia (Wenlock) and Ducie Rivers empty themselves ...'. But what we observed, and what I read from the chart Aus 301, did not seem to invite such confidence today.

65 Report by the Australian Fishing Industry to the Industry Commission, 1993.

66 J. E. Heeres, *The Part Borne by the Dutch* ... pp. 12, 13.

Index

Page numbers referring to illustrations and their captions are listed in italics.